A WORKBOOK IN MATHEMATICAL METHODS FOR ECONOMISTS

R. Quentin Grafton
University of Ottawa

Timothy C. Sargent
University of Ottawa

THE McGRAW-HILL COMPANIES, INC.

New York   St. Louis   San Francisco   Auckland
Bogotá   Caracas   Lisbon   London   Madrid
Mexico City   Milan   Montreal   New Delhi   San Juan
Singapore   Sydney   Tokyo   Toronto

McGraw-Hill
A Division of The **McGraw·Hill** Companies

A WORKBOOK IN MATHEMATICAL METHODS FOR ECONOMISTS

1 2 3 4 5 6 7 8 9 0 QPD QPD 9 0 9 8 7

ISBN 0-07-007122-5

The editor was Lucille Sutton;
the production supervisor was Diane Dilluvio.
The cover was designed by Christopher Brady.
Quebecor-Dubuque was printer and binder.

http://www.mhcollege.com

To Carol-Anne and Anik

Contents

Preface

This workbook is written for all those people who have tried problems in mathematics and felt the frustration of not obtaining the correct answer and not knowing where they went wrong. The workbook is structured on the belief that for most people mathematical techniques can only be learned through the use of examples and by independently working through problems and verifying the solutions. Readers who *first* attempt the problems and then verify their answers will be able to master many of the mathematical methods needed to study intermediate and advanced undergraduate economics.

Our goal is to provide the reader with the resources to become proficient in the mathematical analysis used in economics and business. The self-study approach of the workbook is well-suited to students taking courses in mathematical economics and economic theory and who wish to practise the techniques and methods learned in class. The workbook is also suitable for readers who may have previously learned the topics presented in the workbook but who need a review and practice. To help the reader, each chapter includes a review and introduction of the techniques used in the questions. The review is *not* a substitute to consulting the recommended texts because space limitations prevent the inclusion of almost all the theorems and proofs that underlie the techniques.

We assume the reader has a basic knowledge of algebra and the rules of differentiation. As much as is possible, each chapter is structured to stand alone and, where necessary, references to other chapters are provided. In general, the questions are more difficult at the end than at the beginning of each chapter. We recommend, therefore, that the reader answer the lower numbered problems first so as to build the confidence and skills required for the more advanced questions.

An important feature of the workbook is the emphasis on objective-based learning. To this end, we list the skills that will be acquired by correctly answering the questions given in each chapter. This provides a ready-made structure for self-learning and enables those readers who only wish to master sub-topics within a chapter to consult the relevant questions.

We are grateful for financial support from the University of Ottawa in preparing the manuscript and for the facilities provided by the Norwegian School of Economics and Business Administration (NHH) and the Centre for Fisheries Economics during the summer of 1993. We also thank Stein Ivar Steinshamn of NHH and Chantale Lacasse and David Gray of the University of Ottawa for valuable comments on various chapters. We are especially indebted to Greg Flanagan of Mount Royal College for his encouragement and his many suggestions which have greatly improved the workbook. We would also like to thank students at the University of Ottawa for finding errors in the manuscript and for their many useful questions and comments. The efforts of Milan Jayasinghe and Sean Maguire who provided editorial assistance, Greg Flanagan who helped prepare the camera-ready copy, and Erin Flanagan who prepared the index are much appreciated.

R.Q.G and T.C.S
Ottawa, Canada
August, 1996

Chapter 1

Introduction

Mathematics and Economics

Economics involves the observation of real phenomena, their characterization into a simplified or abstract view of the world and then an analysis and interpretation of potential outcomes. Such a process is called model building by economists and is used to examine positive ("what is") and normative ("what should be") questions about the world around us.

Model building involves abstract representations and can be greatly helped by the language of mathematics. The analysis of economic models may also be improved by the use of mathematical techniques. Thus, a good understanding of mathematics is very useful to economists. Mathematics, however, does not tell economists what are the important economics problems that should be studied nor does it suggest what variables should be the focus of examination or what assumptions are appropriate. Mathematics, therefore, should always be the servant of economics and not *vice versa*. This perspective may seem surprising to the reader who works through the problems in the workbook as the focus is on mathematical technique. Our objective, however, is simply to provide a resource to students so that they can master the mathematical techniques often used in economics. In mastering these techniques we hope that the reader can apply her knowledge and intuition of economics to better address the economic problems that we face.

The exercises in the workbook cover a wide range of applications in economics. Three concepts that underlie the techniques are: Equilibrium; Optimization; Rationality.

Equilibrium. The concept of equilibrium in economics is one of a state of rest where variables will not, in general, change unless there is some perturbation to the system.

Optimization. The notion of optimization is that individuals and firms are economic agents and will always try to do the best they can given their constraints. Thus, if a firm wishes to maximize its profits given a set of market conditions and technology the firm will act in such a way that it does maximize its profits.

Rationality. A related concept is rationality in that economic agents act in a logical and consistent fashion that conforms with their assumed objectives.

All three concepts are a simplification of reality and how people behave but provide the rationale for the use of many of the techniques used by economists. For example, the assumption that agents are rational and optimize is necessary when using calculus to solve for the demand functions of individuals.

1

Outline of the Workbook

This book is not a text in economic theory, economic modeling or rigorous mathematics. It is, no more and no less, a workbook in the mathematical methods used in economics and business. The reader should use the workbook as a supplement to any text used in courses in mathematical economics and for topics in microeconomic and macroeconomic theory. The workbook should be viewed as a resource that can be used to master the mathematical techniques that are an integral part of economic analysis. To this end, the workbook uses objective-based learning to provide structure to the readers self-study of the topics. By consulting the objectives found at the beginning of each chapter, the reader can identify what is to be learned from each question and, if necessary, focus on sub-topics.

Unlike some texts in mathematical methods for economics and business, we actively encourage the reader to consult the many good references on the various topics. At the back of each review section a list of references is provided that should be consulted to understand the theory and proofs behind the techniques and to progess to more advanced topics. In addition, other books are recommended for further practice in problem solving and for applications of the techniques in economics and business. Readers who understand the concepts in the recommended references and correctly answer the questions in the workbook should be confident they have mastered the material.

The workbook follows the topics that are normally covered in the many texts on mathematical methods for economists. Chapter 2 of the workbook uses matrix algebra to examine a number of important economic questions. A large class of economic problems involves systems of linear equations. Such systems of equations can be analyzed and a solution found or characterized using matrices. For example, input-output analysis has been used to examine structural problems in economies and to provide guidance to policy makers where appropriate. Such analysis is greatly helped by a knowledge of matrix algebra. Matrix algebra is also useful in determining whether the solutions to problems are a maximum or a minimum and, thus, are widely used in economics. There are several good texts that can be consulted on matrix algebra. Glaister [19] (chapters 1-10) is particularly useful and provides programs in BASIC to solve problems in matrix algebra. Haeussler and Paul [21] (chapter 6) and Chiang [13] (chapters 4 and 5) are also good references and provide a number of examples and solved problems. We also highly recommend Sydsæter and Hammond [32] (chapters 12-14). Dowling [17] (chapters 10-12) and Toumanoff and Nourzad [34] (chapter 8) also provide many worked examples of how matrices may be applied in economics while Rowcroft [26] (chapter 15) provides an excellent discussion on input-output analysis.

Chapter 3 covers many topics and asks the reader to characterize various functions and sets and use some important theorems and results. The material is fundamental to using mathematics in economics. For example, the ability to identify whether a constraint set is convex or whether the objective function is concave can greatly simplify the solution to a large number of problems. The notion of homogeneity is also widely applied in production and consumer theory and the Taylor series and the implicit function theorem are useful techniques for solving certain economic problems. There are many texts that cover the various topics in the chapter. For the reader who needs a review of algebra and equations before attempting the problems, we recommend Haeussler and Paul [21] (chapters 1 and 2). Dowling [17] (chapters 1 and 2) is also a useful guide to the terminology of mathematics and provides a very good review of differentiation and the use of derivatives in economics (chapters 3 and 4). We also recommend Toumanoff and Nourzad [34] (chapters 3-5) for a good review of differentiation and differentials and their applications to economics. Glaister [19] (chapter 11) provides a good introduction to set theory and its applications in economics while Chiang [13] (chapters 6 and 7), Bressler [10] (chapter 6), Ostrosky and Koch [25] (chapter 3) and Sydsæter and Hammond [32] (chapters 6 and 7) are good references on limits, continuity and differentiation.

The concept of optimization is directly applied in chapters 4-8. Chapter 4 examines the problem of optimization without constraints while chapter 5 addresses this question when an agent's behaviour is constrained. Constrained optimization refers to a wide variety of problems including an individual maximizing her utility subject to a budget constraint or a firm maximizing profits subject to a given production technology. Particularly useful references on unconstrained optimization include Birchenhall and Grout [6] (chapter 5), Chiang [13] (chapter 12), Archibald and Lipsey [1] (chapters 6 and 7), Glaister (chapter 15), Baldani et al. [2] (chapters 7 and 8) and Sydsæter and Hammond [32] (chapter 9 and 17). For constrained optimization we recommend Chiang [13] (chapter 12), Birchenhall and Grout [6] (chapter 8), Archibald and Lipsey [1] (chapters 10 and 11), Baldani et al. (chapters 11 and 12) and Sydsæter and Hammond [32] (chapter 18).

When the solution of constrained optimization problems are substituted into the original objective function of the agent one obtains the optimum value or indirect objective function. An integral part of microeconomics is duality theory which uses the relationship between the optimal value function and the objective function to solve a wide variety of problems. Chapter 6 uses a number of important results from duality theory to solve consumer and producer problems. A valuable reference on duality results and how they may be obtained using the envelope theorem is presented in Silberberg [29] (chapter 7). A useful introduction to duality with worked examples is Birchenhall and Grout [6] (chapters 10-12) while Dixit [15] (chapter 5) and Sydsæter and Hammond [32] (chapter 18.7-18.8) offer a good overview of maximum value functions and the envelope theorem. Baldani et al. [2] also provides a number of applications using duality theory and the envelope theorem (chapter 14).

An optimization problem characterized by an objective function and constraints that are linear in the unknown variables is called linear programming and can be solved using a powerful algorithm, the simplex method. Chapter 7 uses the simplex method to solve for a number problems that arise in business and addresses some of the difficulties when the problem is misspecified. Definitive references on linear programming include Bradley et al. [9], Dorfman et al. [16] and Wu and Coppins [38]. Dorfman et al. [16], in particular, provides many applications of linear programming to economics.

Linear programming is a subset of nonlinear programming where the unknown variables may be nonlinear in the objective function and/or constraints. In reality, most of the optimization problems actually solved by people or firms are nonlinear problems. Chapter 8 uses techniques, such as the Kuhn-Tucker conditions, to solve for demand functions when an individual faces inequality constraints. There are many references on nonlinear programming. We recommend Lambert[24] (chapter 5), Chiang (chapter 21), and Beavis and Dobbs [4] (chapter 2).

Many questions in economics are dynamic or involve optimization over time. A basis for understanding dynamic problems is a knowledge of integral calculus presented in chapter 10. Integration involves finding a primitive function from a derived function. This is useful, for example, in finding the total variable cost function from a marginal cost function or to calculate the area beneath a demand function so as to determine an individual's consumer surplus. Integration is also widely used in econometrics which uses statistics to address economic problems. Useful references on integration include Haeussler and Paul [21] (chapters 16 and 17), Chiang [13] (chapter 13), Sydsæter and Hammond [32] (chapter 10) and Holden and Pearson [23] (chapter 6). Dowling [17] (chapters 16 and 17) also provides numerous worked examples in integral calculus.

Dynamic analysis is examined in chapters 11 and 12 on difference and differential equations. Both techniques are useful in characterizing the time paths of economic variables. These techniques are particularly valuable in macroeconomics which is often concerned with the concept of equilibrium and how variables change over time. Chiang [13] (chapters 14-18), and especially Sydsæter and Hammond [32] (chapters 20 and 21), provide a good introduction to the topics while Dowling [17] (chapters 18-20) provides a number of worked examples. In more complex dynamic problems, the time paths of variables may not always be characterized by real numbers. For this reason, questions are provided in chapter 9 so as to provide the reader with the skills necessary to characterize more complicated problems using complex numbers. A definitive reference on complex numbers is Sydsæter [31] (chapter 2).

To get the most out of the workbook you must *first* attempt the questions before looking at the solutions. It is only by working through problems that you will be able to master the concepts and techniques. It is has been our experience that students who look at the solutions before attempting the questions rarely master the material. Only after you have obtained a solution or are unable to make any further progress on a problem should you consult the solutions. To build your skills progressively, the more difficult questions are located towards the end of each chapter.

There are many good computer programs available that can solve most of the problems in the workbook. We strongly believe, however, that to learn many mathematical techniques it is important to know how to solve problems without the aid of a computer. Thus, we recommend that you first solve the problems in the workbook by hand and then compare your answer with our solutions and the solutions you may obtain using a computer.

Message to the Instructor

The table of contents provides a ready guide to instructors who wish to have their students learn, review and apply particular tools of mathematical analysis. A glossary of over 200 definitions is also provided at the back of the workbook for students wishing to know the meaning of the many terms used in mathematical economics.

The workbook has been used as a supplementary text in a first and second course in mathematical economics. Chapters 2-4 are used in the first course and chapters 5-12 in the second course. Students are assigned problems as the topics are covered in class and are expected to work through the problems. Students of all abilities greatly appreciate the opportunity to work through questions themselves and to correct and learn from their own mistakes. Where necessary, additional problems have also been assigned to provide further practice to the students.

In addition to courses in mathematical economics, chapters 2 and 10 have been assigned to students requiring a review in matrix algebra and integration and who are enrolled in courses in econometrics and statistics. The objectives at the start of the chapter help students identify gaps in their knowledge which can be remedied by doing the problems and consulting the relevant texts.

Other chapters of the book may also be used in second and third courses in microeconomics and macroeconomics. Chapter 5—optimization with equality constraints—was written for and has been used by students in first and second courses in microeconomics. This chapter provides a review of the method of Lagrange and gives students the opportunity to solve economic problems with equality constraints. Chapter 6 on duality theory and chapter 8 on nonlinear programming are particularly useful for students in a second and third course in microeconomics.

Finally, the workbook would also be a valuable text for students taking a review course in mathematical economics in graduate programs in economics. Parts of various chapters should also prove useful in other courses in economics and business.

Chapter 2

Matrix Algebra

Objectives

The questions in this chapter will help the reader to master the fundamentals of matrix algebra. Readers who can correctly answer all the questions should be able to:

1. Determine whether the columns of a matrix are linearly independent (Question 3).

2. Determine the rank of a matrix (Questions 2, 3 and 4).

3. Solve for the inverse of a matrix (Questions 1, 4 and 5).

4. Solve simple problems in input-output analysis (Question 5).

5. Solve systems of linear simultaneous equations (Questions 4, 5 and 6) using elementary row operations, the method of the inverse, and Cramer's rule.

6. Find eigenvalues and eigenvectors (Question 7).

7. Determine the definiteness of a quadratic form (Question 8).

Review

Matrices occur frequently in economics. Many economic models can be expressed as systems of linear equations, and matrix algebra provides efficient methods for solving such problems. Indeed, most of the chapters in the handbook make use of matrix algebra in one form or another.

A matrix consists of numbers and/or variables called *elements* arrayed in a particular way. Suppose, for example, that we have the following system of equations:

$$3x_1 + 4x_2 = -7y_1 + 2y_2$$
$$4x_1 = 9y_2.$$

This system can be written more compactly as

$$\mathbf{Ax} = \mathbf{By}$$

where we define the matrices in upper case letters as $\mathbf{A}$ and $\mathbf{B}$ by

$$\mathbf{A} = \begin{pmatrix} 3 & 4 \\ 4 & 0 \end{pmatrix} \text{ and } \mathbf{B} = \begin{pmatrix} -7 & 2 \\ 0 & 9 \end{pmatrix};$$

and the vectors in lowercase letters $\mathbf{x}$ and $\mathbf{y}$ by

$$\mathbf{x} = \begin{pmatrix} x_1 \\ x_2 \end{pmatrix} \text{ and } \mathbf{y} = \begin{pmatrix} y_1 \\ y_2 \end{pmatrix}.$$

The *elements* of a matrix can be described by the row and column where they are located. Thus, $b_{12} = 2$ represents the element in the first row and second column of the matrix $\mathbf{B}$ while $a_{11} = 3$ is the element in the first row and first column of matrix $\mathbf{A}$. The *dimension* or *order* refers to the number of rows and columns of a matrix. For example, a matrix with 2 rows and 3 columns has the dimension (2×3). The matrices $\mathbf{A}$ and $\mathbf{B}$ have the dimension (2×2).

A *transpose* of a matrix $\mathbf{A}$ is the matrix rewritten such that first, second, $\cdots$ rows become the first, second, $\cdots$ columns. The transpose of a matrix $\mathbf{A}$ is defined as $\mathbf{A}^T$. Thus for the matrix $\mathbf{B}$ defined above, its transpose is:

$$\mathbf{B}^T = \begin{pmatrix} -7 & 0 \\ 2 & 9 \end{pmatrix}$$

A *square* matrix is a matrix where the number of rows equals the number of columns. The *main diagonal* of a square matrix is the elements formed by drawing a line from the upper left corner to the bottom right corner. Thus, the main diagonal of the matrix $\mathbf{A}$ above are the elements 3 and 0. An *upper triangular* matrix is a square matrix where all the elements to the left of the main diagonal equal zero. A *lower triangular* matrix is a square matrix where all the elements to the right of the main diagonal are zero. A *symmetric* matrix is a square matrix where $a_{ij} = a_{ji}$ for $i \neq j$.

A vector is a matrix with only one row or column. The *length* of a vector $\mathbf{x}$ is denoted by $|\mathbf{x}|$ and is computed using the *Euclidean formula*

$$|\mathbf{x}| \;=\; \sqrt{x_1^2 + x_2^2 + \cdots x_n^2}$$

where n is the dimension of the vector. A vector of unit length has a length of 1.

A set of vectors is said to be *linearly dependent* if one of the vectors can be expressed as a linear combination of the others. For example, consider the vectors $\mathbf{v}_1$ and $\mathbf{v}_2$, where:

$$\mathbf{v}_1 = \begin{pmatrix} 4 \\ 2 \end{pmatrix}, \mathbf{v}_2 = \begin{pmatrix} 1 \\ 0.5 \end{pmatrix}$$

These two vectors are linearly dependent because $\mathbf{v}_1 = 4\mathbf{v}_2$.

More formally, a set of m vectors $\mathbf{v}_1,...,\mathbf{v}_m$ is linearly dependent if $\sum_{i=1}^{m} c_i \mathbf{v}_i = 0$ and $c_i \neq 0$ for at least one c_i. A set of vectors is *linearly independent* if none of the vectors can be expressed as a linear combination of the others, so that $\sum_{i=1}^{m} c_i \mathbf{v}_i = 0$ is *only* true when each scalar $c_i = 0$.

A useful way to test for linear dependence is to write the vectors as a system of linear equations in c_i. Consider the vectors $\mathbf{v}_1$ and $\mathbf{v}_2$ above. They will be linearly dependent if there exists $(c_1, c_2) \neq (0, 0)$ such that:

$$c_1 \begin{pmatrix} 4 \\ 2 \end{pmatrix} + c_2 \begin{pmatrix} 1 \\ 0.5 \end{pmatrix} = \begin{pmatrix} 0 \\ 0 \end{pmatrix}.$$

This condition can be restated as the following two equations:

$$4c_1 + c_2 = 0 \Rightarrow 4c_1 = -c_2$$
$$2c_1 + 0.5c_2 = 0 \Rightarrow 2c_1 = -0.5c_2$$

Substituting the second equation into the first we obtain

$$c_2 = c_2$$

This is true for any value of c_2 and not just $c_2 = 0$. Thus, the vectors $\mathbf{v_1}$ and $\mathbf{v_2}$ are linearly dependent.

Two vectors are *orthogonal* if the vectors are perpendicular (at right angles) to each other. Formally, this definition requires that $\mathbf{v}_1^T \cdot \mathbf{v}_2 = 0$, where the dot operator '·' is defined as the *inner product*: the sum of the product of the corresponding elements in each vector. Using the definitions of $\mathbf{v}_1$ and $\mathbf{v}_2$ from above, we find that

$$\mathbf{v}_1^T \cdot \mathbf{v}_2 = \begin{pmatrix} 4 & 2 \end{pmatrix} \begin{pmatrix} 1 \\ 0.5 \end{pmatrix} = (4 \times 1) + (2 \times 0.5) = 5$$

and the two vectors are therefore not orthogonal. A set of vectors which are mutually orthogonal and where each vector is of unit length is an *orthonormal* set. A vector can be *normalized* to be of unit length by dividing each element of the vector by its length.

Matrices can be added and subtracted provided that they have the same dimension. Note that it is not necessary that the matrices be square. The procedure for addition (subtraction) is to add (subtract) the corresponding element in the second matrix to (from) the first matrix. For example, consider the matrices:

$$\mathbf{C} = \begin{pmatrix} 12 & 5 & -3 \\ 0 & 2 & 1 \end{pmatrix}, \mathbf{D} = \begin{pmatrix} 6 & 6 & -2 \\ -1 & 0 & 1 \end{pmatrix}.$$

Then

$$\mathbf{C} - \mathbf{D} = \begin{pmatrix} 12-6 & 5-6 & (-3)-(-2) \\ 0-(-1) & 2-0 & 1-1 \end{pmatrix} = \begin{pmatrix} 6 & -1 & -1 \\ 1 & 2 & 0 \end{pmatrix};$$

Matrices can be multiplied by a number, in matrix terminology, a scalar. In this case, every element in the matrix is multiplied by this number. Matrices can also be multiplied by another matrix provided the two matrices are *conformable*, i.e., that the number of columns in the premultiplying matrix equals the number of rows in the postmultiplying matrix. The product of the two matrices has a dimension equal to the number of rows in the premultiplying matrix and the number of columns in the postmultiplying matrix. Matrix multiplication is illustrated as follows where $\mathbf{A}$ premultiplies $\mathbf{D}$:

$$\mathbf{AD} = \begin{pmatrix} (3 \times 6)+(4 \times -1) & (3 \times 6)+(4 \times 0) & (3 \times -2)+(4 \times 1) \\ (4 \times 6)+(0 \times -1) & (4 \times 6)+(0 \times 0) & (4 \times -2)+(4 \times 1) \end{pmatrix} = \begin{pmatrix} 14 & 18 & -2 \\ 24 & 24 & -8 \end{pmatrix};$$

Thus the first element in new matrix is the sum of the product of the elements in the first row of matrix $\mathbf{A}$ multiplied by the corresponding elements in the first column of the matrix $\mathbf{B}$. Matrix multiplication does not, in general, satisfy the commutative property of multiplication, so that the order of multiplication matters. It does, however, satisfy the distributive property $\mathbf{A(B + C) = AB + AC}$ and the associative property $\mathbf{A(BC) = (AB)C}$.

An important concept in matrix algebra which applies only to square matrices is the *determinant*. The determinant of a matrix $\mathbf{A}$ is a single number, and it is written as either $|\mathbf{A}|$ or $\det \mathbf{A}$. For a matrix with only one element the determinant is the element itself. For a square matrix of dimension 2, the determinant is the product of the elements of the main diagonal less the product of the other two elements. Thus, for matrix $\mathbf{B}$ above the determinant is:

$$|\mathbf{B}| = \det \mathbf{B} = (-7 \times 9) - (0 \times 2) = -63$$

The determinant of a matrix $\mathbf{A}$ is the same as the determinant of its transpose $\mathbf{A}^T$ and the determinant of a matrix $\mathbf{A}$ of dimension n multiplied by a scalar α is $\alpha^n |\mathbf{A}|$. For upper and lower triangular matrices the determinant is the product of the elements on the main diagonal.

Calculating the determinant of a higher order square matrices is somewhat more complicated than calculating the determinant of a 2×2 matrix. The technique we use here is known as the *Laplace expansion*: it uses the fact that the determinant of a matrix is the sum of the elements of any row or column multiplied by their respective *cofactors*. The cofactor of the element a_{ij} of a matrix $\mathbf{A}$ is defined as c_{ij} and is the product of $(-1)^{i+j}$ and the *minor* of a_{ij}. The minor of an element a_{ij} is defined as the determinant of the matrix M_{ij} that is formed by deleting the ith row and jth column. The definition is applied in the following example:

$$\mathbf{C} = \begin{pmatrix} 1 & 0 & 4 \\ 6 & 1 & 3 \\ 5 & 3 & 1 \end{pmatrix}.$$

We choose the first row to calculate the determinant because it contains a zero element which simplifies our calculations. The first task is to calculate the matrices M_{ij} corrsponding to each of the three elements in the first row. These are

$$M_{11} = \begin{pmatrix} 1 & 3 \\ 3 & 1 \end{pmatrix},$$

$$M_{12} = \begin{pmatrix} 6 & 3 \\ 5 & 1 \end{pmatrix},$$

$$M_{13} = \begin{pmatrix} 6 & 1 \\ 5 & 3 \end{pmatrix}.$$

The cofactors of the elements of the top row are then given by

$$c_{11} = (-1)^{1+1} \det \begin{pmatrix} 1 & 3 \\ 3 & 1 \end{pmatrix},$$

$$c_{12} = (-1)^{1+2} \det \begin{pmatrix} 6 & 3 \\ 5 & 1 \end{pmatrix},$$

$$c_{13} = (-1)^{1+3} \det \begin{pmatrix} 6 & 1 \\ 5 & 3 \end{pmatrix}.$$

The sum of each element in the first row multiplied by its cofactor is then

$$
\begin{aligned}
\det(\mathbf{C}) &= (1)c_{11} + (0)c_{12} + (4)c_{13} \\
&= (1)(-1)^{1+1} \det \begin{pmatrix} 1 & 3 \\ 3 & 1 \end{pmatrix} + (0)(-1)^{1+2} \det \begin{pmatrix} 6 & 3 \\ 5 & 1 \end{pmatrix} \\
&\quad + (4)(-1)^{1+3} \det \begin{pmatrix} 6 & 1 \\ 5 & 3 \end{pmatrix} \\
&= (1)(1)(-8) + (0)(-1)(-9) + (4)(1)(13)
\end{aligned}
$$

which solves to

$$
\det(\mathbf{C}) = -8 + 0 + 52 = 44.
$$

The concept of *definiteness* is used to check the characteristics of functions (see chapter 3), and it is particularly useful in solving nonlinear optimization problems. The quadratic form of a two variable function is defined as

$$
Q(x) = ax_1^2 + 2bx_1x_2 + cx_2^2
$$

For an n variable function the quadratic form is

$$
Q(x) = a_{11}x_1^2 + a_{12}x_1x_2 + \cdots + a_{ij}x_ix_j + \cdots + a_{nn}x_n^2
$$

Using matrices, the quadratic form is defined as

$$
Q(x) = \mathbf{x}^T \mathbf{A} \mathbf{x}
$$

where $\mathbf{x}$ is a column vector of the x variables and the matrix $\mathbf{A}$ is a square and symmetric matrix defined below

$$
\mathbf{A} = \begin{pmatrix} a_{11} & \cdots & a_{1n} \\ \vdots & & \vdots \\ a_{n1} & \cdots & a_{nn} \end{pmatrix}.
$$

If Q is always positive (negative) for all values of x where $(x_1, x_2 \cdots x_n) \neq (0, 0 \cdots 0)$ then the matrix A is positive (negative) definite. If Q is always equal to or greater (less) than 0 for all values of x where $(x_1, x_2, \cdots x_n) \neq (0, 0, \cdots 0)$ then $\mathbf{A}$ is positive (negative) semidefinite.

To check the definiteness of the $\mathbf{A}$ matrix we determine the sign of the successive leading principal minors which are the determinants of the square submatrices formed from the main diagonal starting with the element in the first row and first column followed by the elements in the first row and first column and the second row and second column, and so on. Thus, for the three by three matrix

$$
\mathbf{A} = \begin{pmatrix} a_{11} & a_{12} & a_{13} \\ a_{21} & a_{22} & a_{23} \\ a_{31} & a_{23} & a_{33} \end{pmatrix},
$$

the successive leading principal minors are

$$\det(a_{11}) = a_{11},$$

$$\det \begin{pmatrix} a_{11} & a_{12} \\ a_{21} & a_{22} \end{pmatrix},$$

and

$$\det(\mathbf{A}) = \det \begin{pmatrix} a_{11} & a_{12} & a_{13} \\ a_{21} & a_{22} & a_{23} \\ a_{31} & a_{23} & a_{33} \end{pmatrix}.$$

In general, a matrix $\mathbf{A}$ is positive definite if each successive leading principal minor is positive and each element on the main diagonal is positive. It is negative definite if the first leading principal minor $(\mathbf{a}_{11})$ is negative, the next is positive, the next is negative, and so on. More formally, negative definiteness requires that $(-1)^n|\mathbf{D}_n| > 0$ where $|\mathbf{D}_1|$ is the first leading principal minor and $|\mathbf{D}_n|$ is the nth leading principal minor. The $\mathbf{A}$ matrix may also be positive and negative *semidefinite*. To test semidefiniteness *all* the principal minors and not just the leading principal minors must be tested. For negative semidefiniteness *all* the principal minors of order n obtained from a matrix of order n defined as $|\mathbf{B}_n|$ must satisfy $(-1)^n|\mathbf{B}_n| \geq 0$ while for positive semidefiniteness all the principal minors ≥ 0. The principal minors are the determinants of the submatrices formed by deleting k columns and k rows from the original matrix where $k < n$ and n is the dimension of the matrix. Thus, for a 2×2 matrix the principal minors include the determinant of the matrix itself and the elements on the main diagonal. A matrix which is neither positive nor negative semidefinite is said to be *indefinite*.

The *inverse* of a square matrix $\mathbf{A}$ is defined as $\mathbf{A}^{-1}$ and when pre- or post-multiplied by $\mathbf{A}$ yields the *identity* matrix $\mathbf{I}$. Thus,

$$\mathbf{A}\mathbf{A}^{-1} = \mathbf{A}^{-1}\mathbf{A} = \mathbf{I}$$

The identity matrix is a square matrix where all the elements on the main diagonal equal one and all other elements are zero and can be any dimension or order n where $n \geq 2$ is an integer. Thus, $\mathbf{I}_3$ is defined as:

$$\mathbf{I}_3 = \begin{pmatrix} 1 & 0 & 0 \\ 0 & 1 & 0 \\ 0 & 0 & 1 \end{pmatrix};$$

The identity matrix acts like the number 1 in the multiplication of real numbers in that for a square matrix $\mathbf{X}$ of order n, $\mathbf{I}_n\mathbf{X} = \mathbf{X}\mathbf{I}_n = \mathbf{X}$. The identity matrix is also an *idempotent* matrix in that the identity matrix multiplied by itself gives the identity matrix, i.e., $\mathbf{I}_n\mathbf{I}_n = \mathbf{I}_n$.

The inverse of a square matrix exists only if the determinant does not equal zero. Such matrices are called *invertible* or *non-singular*. If the determinant of a matrix equals zero then it is a *singular* matrix. The inverse is calculated using the determinant and the *adjoint* matrix where the adjoint matrix is the transpose of the matrix of cofactors. For the matrix $\mathbf{A}$, the inverse is defined as

$$\mathbf{A}^{-1} = \frac{\mathrm{adj}(\mathbf{A})}{\det(\mathbf{A})}$$

Thus for the matrix

$$\mathbf{A} = \begin{pmatrix} a & b \\ c & d \end{pmatrix}$$

the inverse is

$$\mathbf{A}^{-1} = \frac{1}{(ad - bc)} \begin{pmatrix} d & -b \\ -c & a \end{pmatrix};$$

The inverse may also be calculated using elementary row operations which consist of interchanging of rows, adding a multiple of one row to another, and multiplying a row by a non-zero scalar. To find the inverse of a matrix $\mathbf{A}$ set up a new matrix $[\mathbf{A}|\mathbf{I_n}]$ and use elementary row operations to transform the matrix to $[\mathbf{I_n}|\mathbf{B}]$ where $\mathbf{B}$ will be the inverse of $\mathbf{A}$. Two useful rules of inverses are one, if a matrix is invertible so is its transpose and two, if two matrices $\mathbf{A}$ and $\mathbf{B}$ are invertible then $(\mathbf{AB})^{-1} = \mathbf{B}^{-1}\mathbf{A}^{-1}$.

An important application of inverses in economics is *input-output analysis*. Developed independently by P. Sraffa and W. Leontief, input-output analysis is used to examine demand and supply relationships in various sectors and industries of an economy at a particular point in time. It can be used to determine "bottlenecks" and to predict how the outputs of different sectors must change to satisfy a change in total consumer expenditure. An input-output table is an $n \times n$ matrix that has as its coefficients the input produced by an industry for its own output and its input for other industries. Often when presenting an input-output table, the costs of various productive factors, such as wages and profits, are also included, along with the exogenously determined final consumer demands. The total payments from industry to households in the form of wages and profits equals the sum of the final demands. A simplified input-output table which includes the contribution of other productive factors (labor) is presented below.

	Agriculture	Manufacturing
Agriculture	0.5	0.1
Manufacturing	0.3	0.6
Labor	0.2	0.3

The elements in the input-output table are called *technical coefficients*, where a_{ij} represents the value of input i to produce one dollar's worth of output j and $0 \le a_{ij} \le 1$. For example, for every dollar worth of output produced in agriculture 50 cents of input is required from agriculture, 30 cents of input is required from manufacturing and 20 cents is spent on labor. Thus, each column represents the input cost incurred in producing a dollar's worth of output in that sector.

To determine how changes in final demands affect the outputs in each sector we can use the *Leontief matrix*. To create the Leontief matrix we define $\mathbf{A}$ as a matrix of technical coefficients obtained from the input-output table by deleting the row which gives the contribution to other factors. Thus:

$$\mathbf{A} = \begin{pmatrix} 0.5 & 0.1 \\ 0.3 & 0.6 \end{pmatrix}$$

In addition, we define $\mathbf{x}$ as a column vector of the total value of each sector and $\mathbf{d}$ as a column vector giving the final demands of each sector. For an exogenously given $\mathbf{d}$ we can calculate the total value of the outputs in each sector as follows:

$$\mathbf{x} = (\mathbf{I} - \mathbf{A})^{-1}\mathbf{d}$$

where $\mathbf{I} - \mathbf{A}$ is the Leontief matrix.

If $\det(\mathbf{I} - \mathbf{A}) = 0$ then the inverse does not exist and the system cannot satisfy any positive set of final demands. The conditions that ensure feasibility of an n sector economy are called the *Hawkins-Simon* condition and require that all the elements on the main diagonal be positive and that all the leading principal minors of the ($\mathbf{I} - \mathbf{A}$) matrix be positive.

The *rank* of a matrix equals the number of linearly independent columns. The rank of any matrix equals the dimension or order of the largest square matrix with a non-zero determinant which can be formed by deleting rows, columns, or both. A square matrix has *full rank* if its rank equals the order of the matrix itself. The concept of rank does not only apply to square matrices. The rank of a matrix can never exceed the minimum of the number of rows or columns.

When solving a system of simultaneous equations, the concept of rank can be used to determine if a solution exists. For the system of simultaneous equations defined as $\mathbf{Ax} = \mathbf{b}$ where $\mathbf{A}$ is a matrix of coefficients, $\mathbf{x}$ is a vector of unknown variables and $\mathbf{b}$ is a vector of known constants a solution (not necessarily unique) exists **if and only if** rank($\mathbf{A}$)=rank($\mathbf{A}|\mathbf{b}$). Where a solution exists the system of equations is *consistent*.

A system of linear equations defined by $\mathbf{Ax} = \mathbf{b}$ where $\mathbf{b}$ is a *null* column vector which consists entirely of zeros is called a *homogeneous* system. For a homogenous system of linear equations to have a *non-trivial* solution such that $\mathbf{x}$ is **not** a null vector, it must be true that the *coefficient matrix*—the matrix of known coefficients—is singular, so that $\det(\mathbf{A}) = 0$.

When solving systems of simultaneous linear equations we may use elementary row operations in the method of reduction, the inverse of a matrix or *Cramer's rule*. The method of reduction uses elementary row operations to derive a *reduced* matrix from the *augmented coefficient* matrix defined as $[\mathbf{A}|\mathbf{b}]$ for the system of simultaneous equations defined by $\mathbf{Ax} = \mathbf{b}$. The elementary row operations include one or a combination of the following operations.

1. Any rows can be interchanged, i.e., $R_i \Leftrightarrow R_j$.

2. Rows may be multiplied by a non-zero scalar, i.e., cR_i.

3. Rows and constant multiples of rows can be added or subtracted, i.e., $cR_i + R_j$.

While a reduced matrix has the following properties

1. Provided the row does not contain all zeros the first or leading element must equal 1.

2. The leading element of each row is to the right of the leading element of the row above.

3. All zero rows (where all elements are zero) are at the bottom of the matrix.

The reduced matrix can also be used to determine the rank of $\mathbf{A}$ and the rank of $[\mathbf{A}|\mathbf{b}]$. The rank of $\mathbf{A}$ is the number of rows which do not contain all zeros on the left hand side of the reduced matrix, not including the last column, and the rank of $[\mathbf{A}|\mathbf{b}]$ is the number of rows which do not contain all zeros in the full reduced matrix.

If n is the number of unknowns and m is the number of equations for a system of linear equations then if $n = m$ there are three possibilities.

1. If rank($\mathbf{A}$)=rank($\mathbf{A}|\mathbf{b}$)=n then the system is of *full rank*, consistent and there is a unique solution.

2. If rank($\mathbf{A}$)=rank($\mathbf{A}|\mathbf{b}$)< n then the system is consistent but the solution is not unique.

3. If rank($\mathbf{A}$)≠rank($\mathbf{A}|\mathbf{b}$)=n then the system is inconsistent.

Where $n > m$ then,

1. If rank($\mathbf{A}$)=rank($\mathbf{A}|\mathbf{b}$) then the system is consistent but the solution is not unique.

2. If rank($\mathbf{A}$)≠rank($\mathbf{A}|\mathbf{b}$)=n then the system is inconsistent.

Where $n < m$ then,

1. If rank($\mathbf{A}$)=rank($\mathbf{A}|\mathbf{b}$)=n then the system is consistent and there is a unique solution.

2. If rank($\mathbf{A}$)= n ≠rank($\mathbf{A}|\mathbf{b}$) then the system is inconsistent.

3. If rank($\mathbf{A}$)=rank($\mathbf{A}|\mathbf{b}$)< n then the system is consistent but the solution is not unique.

4. If rank($\mathbf{A}$)≠rank($\mathbf{A}|\mathbf{b}$) then the system is inconsistent.

Unlike the method of elementary row operations, the inverse of a matrix and Cramer's rule can only be used to solve systems of simultaneous linear equations where the number of unknowns equals the number of equations. Using the inverse, the solution to a system of simultaneous equations defined by $\mathbf{Ax} = \mathbf{b}$ is the value of the elements of the unknown vector $\mathbf{x}$. The solution is defined as $\mathbf{x} = \mathbf{A}^{-1}\mathbf{b}$ and exists *if and only if* det($\mathbf{A}$) $\neq 0$.

Cramer's rule is another method for solving for the unknowns. Provided that $\mathbf{A}^{-1}$ exists (det $\mathbf{A} \neq 0$) then the ith element of the vector of unknowns $\mathbf{x}$ is calculated as follows $\mathbf{x_i} = \frac{\det \mathbf{B}_i}{\det \mathbf{A}}$ where the matrix $\mathbf{B_i}$ is the matrix $\mathbf{A}$ with the ith column replaced by the vector $\mathbf{b}$. Cramer's rule is particularly useful when we are only interested in knowing a subset of the unknowns.

Another important topic in matrix algebra concerns equations of the form:

$$\mathbf{Ax} = \lambda \mathbf{x}$$

where $\mathbf{x}$ is a non-zero vector of unknowns. Expressions like this occur frequently when solving systems of differential equations (see chapter 12). The values of λ and $\mathbf{x}$ that solve the above are known respectively as the eigenvalues and eigenvectors of the system. The above expression may be rewritten as $(\mathbf{A} - \lambda\mathbf{I})\mathbf{x} = \mathbf{0}$. This, in turn, requires that the matrix $(\mathbf{A} - \lambda\mathbf{I})$ be singular or that det$(\mathbf{A} - \lambda\mathbf{I}) = 0$. Using this result, the eigenvalues can be determined by calculating det$(\mathbf{A} - \lambda\mathbf{I})$ and deriving an expression in terms of λ. This expression, known as the *characteristic equation*, can then be solved to determine the eigenvalues or the values of λ that make the expression equal to zero. The eigenvectors are calculated by substituting each eigenvalue into $(\mathbf{A} - \lambda\mathbf{I})\mathbf{x} = \mathbf{0}$ and then obtaining an expression in terms of x. Not infrequently when solving for the eigenvalues of a matrix of order 2 we need to use the *quadratic formula*, which states that the solutions to a quadratic equation of the form:

$$ax^2 + bx + c = 0, (a \neq 0)$$

can be computed using the quadratic formula

$$x = \frac{-b \pm \sqrt{b^2 - 4ac}}{2a}$$

Eigenvalues can also be used to test the definiteness of a matrix. If $\mathbf{A}$ is a *symmetric* matrix, then if all the eigenvalues are positive (negative), the matrix is positive (negative) definite. If at least two eigenvalues have a different sign then the matrix is indefinite and if all the eigenvalues are equal to or greater (less) than 0 the matrix is positive (negative) semidefinite.

To conclude the review section, we should point out that the there are many software packages that quickly calculate inverses, determinants and eigenvalues, and many other operations in matrix algebra. Packages such as Mathematica and Maple can even perform symbolic matrix algebra: We can specify a matrix in terms of symbols such as x, y, and z, and receive an answer in terms of these symbols. However, in the same way that the invention of pocket calculators has not freed us from the need to do simple arithmetic, mathematical software programs have not eliminated the need for us to do matrix algebra manually, at least for small matrices. The ability to perform such computations is important if we are to grasp the **concepts** behind matrix algebra.

Further Reading

Textbooks in mathematical economics that include readable introductions to matrix algebra are Chiang [13] (chapters 4 and 5), Archibald and Lipsey [1] (chapter 4), Bressler [10] (chapter 5), Holden and Pearson [23] (chapter 2), Haeussler and Paul [21] (chapter 6), Toumanoff and Nourzad [34] (chapter 8) and Glaister [19] (chapters 1-8). Glaister also covers eigenvalues and quadratic forms (chapters 9-10). Sydsæter and Hammond [32] (chapter 13 and 14) is also highly recommended and includes a very good discussion on eigenvalues (chapter 14.4), and the rank of a matrix and solving simultaneous linear equations (chapter 14.2 and 14.3). An excellent review of input-output models is presented in Rowcroft [26] (chapter 15). Rowcroft [26] also provides an application of eigenvalues in economics (chapter 19.2). Additional problems in matrix algebra, with worked examples, are presented in Dowling [17] (chapters 10-12).

A more advanced treatment of the relationship between matrix analysis and input-output models can be found in Takayama [33] (chapter 4) or Woods [37]. A good general reference for the theory of matrix algebra is Strang [30]: although aimed at students in mathematics, the text is quite readable and the emphasis is on applications rather than proofs of theorems.

Chapter 2 - Questions

Matrix Manipulation

Question 1

Suppose that $\mathbf{y} = \begin{pmatrix} 5 \\ 3 \\ 9 \end{pmatrix}$ and $\mathbf{X} = \begin{pmatrix} 1 & 3 & 5 \\ 1 & 2 & 2 \\ 1 & 2 & 3 \end{pmatrix}$.

Compute the following:
(i) $\mathbf{X}^T\mathbf{X}$.

(ii) $\det(\mathbf{X}^T\mathbf{X})$

(iii) $(\mathbf{X}^T\mathbf{X})^{-1}$.

(iv) $\mathbf{X}^T\mathbf{y}$ and thus $\mathbf{b} = (\mathbf{X}^T\mathbf{X})^{-1}\mathbf{X}^T\mathbf{y}$.

(v) Show that $\mathbf{b} = \mathbf{X}^{-1}\mathbf{y}$.

Rank

Question 2

Determine the rank of the following matrices.

(i)

$$\mathbf{A} = \begin{pmatrix} 1 & 2 & -4 \\ 3 & 6 & -12 \\ 4 & 8 & -16 \end{pmatrix}.$$

(ii)

$$\mathbf{A} = \begin{pmatrix} 0 & 0 \\ 0 & 0 \end{pmatrix}.$$

(iii)

$$\mathbf{A} = \begin{pmatrix} 1 & 0 & 4 & 2 \\ 3 & 2 & 1 & 0 \\ 2 & 1 & 3 & 4 \end{pmatrix}.$$

Linear Independence and Rank

Question 3

Let the vectors $\mathbf{v}_1, \mathbf{v}_2$ and $\mathbf{v}_3$ be given by

$$\mathbf{v}_1 = \begin{pmatrix} 5 \\ 1 \end{pmatrix}, \mathbf{v}_2 = \begin{pmatrix} -7 \\ 2 \end{pmatrix} \text{ and } \mathbf{v}_3 = \begin{pmatrix} -10 \\ -2 \end{pmatrix}.$$

(i) Are $\mathbf{v}_1$ and $\mathbf{v}_2$ linearly independent?

(ii) Are $\mathbf{v}_1$ and $\mathbf{v}_3$ linearly independent?

(iii) What is the rank of the matrix $\mathbf{V} = [\mathbf{v}_1 : \mathbf{v}_2 : \mathbf{v}_3]$?

Solving Systems of Linear Equations

Question 4

Determine whether the following linear simultaneous equation systems have unique solutions and if so solve for them.

(i)

$$\begin{aligned} x_1 + x_2 \quad\quad + x_4 &= 1, \\ x_1 \quad\quad + x_3 + x_4 &= 2, \\ 2x_1 + x_2 + x_3 + 2x_4 &= 4. \end{aligned}$$

(ii)

$$\begin{aligned} 3x_1 + 4x_2 &= 4 \\ 6x_1 + 8x_2 &= 8 \end{aligned}$$

(iii)

$$\begin{aligned} x_1 + x_2 + x_3 &= 2 \\ -x_1 + 2x_2 - x_3 &= 4 \\ -2x_1 + x_2 - x_3 &= 6 \end{aligned}$$

Input-Output Analysis

Question 5

Suppose that the technical coefficients for a two sector economy are given below

	Coal	Steel
Coal	0.25	0.1
Steel	0.5	0.4

and that

where the a_{ij}th element represents the contribution of input i to a dollar's worth of output in sector j.

(i) Verify the Hawkins-Simon conditions for this economy: that is, ensure that it is possible to produce a positive vector of *net* output. (In this context we define a positive vector as one where every element is non-negative, and at least one element is strictly positive).

(ii) Suppose that it is decided that final consumption in this economy should be $\mathbf{d} = (50, 50)$: what payment to labor would be required if every dollar of output of coal needs a labor input of 25 cents and every dollar of output of steel needs 50 cents of labor input?

(iii) Suppose the total wage bill in the country can be no more than 100. What is the final consumption possibility set? Is the final demand vector (20,20) feasible in this economy?

National Income Model

Question 6

The following equations describe a macroeconomic model, where the money supply (M) and government expenditure (G) are the exogenous variables which are given; and consumption (C), investment (I), interest rates (R) and output (Y) are the endogenous variables which must be solved for:

$$
\begin{aligned}
Y &= C + I + G \\
C &= \alpha + \beta(1 - \tau)Y - \gamma R \\
I &= \delta + \epsilon(1 - \tau)Y - \zeta R \\
M &= \theta(1 - \tau)Y - \eta R
\end{aligned}
$$

All the parameters are positive, and the tax rate τ is constrained to be less than one.

Solve this model for output (Y) in terms of the parameters ($\alpha, \beta, \tau, \gamma, \delta, \epsilon, \zeta, \theta, \eta$) and the exogenous variables (G and M).

Eigenvalues

Question 7

Consider the following matrices.

$$\mathbf{A}_1 = \begin{pmatrix} 4 & -5 \\ 2 & -3 \end{pmatrix},$$

$$\mathbf{A}_2 = \begin{pmatrix} 1 & 2 \\ -1 & 3 \end{pmatrix},$$

$$\mathbf{A}_3 = \begin{pmatrix} 1 & 0 & 2 \\ 0 & 1 & 0 \\ 2 & 0 & 1 \end{pmatrix}.$$

(i) Find the eigenvalues for each matrix.

(ii) For each matrix with *real* eigenvalues, find the set of eigenvectors.

(iii) Are any of the sets of eigenvectors orthonormal?

Quadratic Forms

Question 8

Put the following functions into their quadratic forms, and determine the definiteness of these quadratic forms.

(i) $f(x_1, x_2) = x_1^2 + 2x_1x_2 + x_2^2$.

(ii) $f(x_1, x_2) = x_1^2 - 2x_1x_2 + x_2^2$.

(iii) $f(x_1, x_2) = -2x_1^2 - 2x_1x_2 - x_2^2$.

(iv) $f(x_1, x_2, x_3) = 2x_1^2 - 2x_1x_2 + x_2^2 - x_2x_3 + 2x_3^2$.

Chapter 2 - Solutions

Matrix Manipulation

Question 1

(i)

$\mathbf{X}^T$ is the *transpose* matrix of $\mathbf{X}$, wherein each row of $\mathbf{X}$ becomes the corresponding column of $\mathbf{X}^T$, so

$$\mathbf{X}^T = \begin{pmatrix} 1 & 1 & 1 \\ 3 & 2 & 2 \\ 5 & 2 & 3 \end{pmatrix}.$$

Before proceeding we must first verify that the number of columns of the first matrix $\mathbf{X}^T$ is the same as the number of rows of the second matrix, $\mathbf{X}$. This is indeed the case, since both matrices have three rows and three columns. Since $\mathbf{X}^T$ has three rows and $\mathbf{X}$ has three columns we can also tell that $\mathbf{X}^T\mathbf{X}$ will also be a three by three matrix.

To proceed with the actual multiplication of the two matrices $\mathbf{X}^T$ and $\mathbf{X}$ we multiply each row of $\mathbf{X}^T$ by each column of $\mathbf{X}$. This is done by multiplying each element in the row by the corresponding element in the column and then summing the products. The number so obtained will be the same row number as the row in $\mathbf{X}^T$ and the same column number as the column in $\mathbf{X}$. Thus to obtain the first element of the second row in $\mathbf{X}^T\mathbf{X}$ we must multiply the second row of $\mathbf{X}^T$ by the first column of $\mathbf{X}$:

$$\begin{pmatrix} 3 & 2 & 2 \end{pmatrix} \cdot \begin{pmatrix} 1 \\ 1 \\ 1 \end{pmatrix} = (3)(1) + (2)(1) + (2)(1) = 7.$$

By applying this methodology for all of the elements of $\mathbf{X}^T\mathbf{X}$ we obtain

$$\mathbf{X}^T\mathbf{X} = \begin{pmatrix} 1 & 1 & 1 \\ 3 & 2 & 2 \\ 5 & 2 & 3 \end{pmatrix} \cdot \begin{pmatrix} 1 & 3 & 5 \\ 1 & 2 & 2 \\ 1 & 2 & 3 \end{pmatrix} = \begin{pmatrix} 3 & 7 & 10 \\ 7 & 17 & 25 \\ 10 & 25 & 38 \end{pmatrix}$$

(ii)

The *determinant* of a matrix $\mathbf{A}$ is a combination of the elements of row i of $\mathbf{A}, \mathbf{a_{ij}}$, and the cofactors of this row C_{ij}:

$$\det(\mathbf{A}) = \mathbf{a}_{i1}C_{i1} + \mathbf{a}_{i2}C_{i2} + \ldots + \mathbf{a_{in}}C_{\mathrm{in}},$$

where the cofactor is computed according to

$$C_{ij} = (-1)^{i+j}\det(M_{ij}),$$

and M_{ij} is the matrix formed by deleting row i and column j from $\mathbf{A}$. Taking the first row, we have

$$\det(\mathbf{X}^T\mathbf{X}) = (3)(-1)^{1+1}\det\begin{pmatrix} 17 & 25 \\ 25 & 38 \end{pmatrix} + (7)(-1)^{1+2}\det\begin{pmatrix} 7 & 25 \\ 10 & 38 \end{pmatrix}$$

$$+(10)(-1)^{1+3}\det\begin{pmatrix} 7 & 17 \\ 10 & 25 \end{pmatrix}.$$

To compute this expression we use the fact that the determinant of a two by two matrix

$$\mathbf{A} = \begin{pmatrix} a_{11} & a_{12} \\ a_{21} & a_{22} \end{pmatrix}$$

is the product of the main diagonal less the product of the other two elements:

$$\det(\mathbf{A}) = a_{11}a_{22} - a_{21}a_{12}.$$

Armed with this formula we can now go on to evaluate the determinant of $\mathbf{X}^T\mathbf{X}$:

$$\det(\mathbf{X}^T\mathbf{X}) = (3)(1)(21) + (7)(-1)(16) + (10)(1)(5) = 1.$$

(iii)

Because the determinant of $\mathbf{X}^T\mathbf{X}$ is one, the inverse is simply the adjoint matrix or the transpose of the matrix of cofactors. Thus, for each row i and each column j we compute the cofactor C_{ij}, the formula for which is given in (ii) above. Applying this formula, we have

$$(\mathbf{X}^T\mathbf{X})^{-1} = \begin{pmatrix} 21 & -16 & 5 \\ -16 & 14 & -5 \\ 5 & -5 & 2 \end{pmatrix}$$

Another method to calculate the inverse that is often easier for matrices of dimension three or larger uses elementary row operations, i.e.,

1. Any rows can be interchanged, i.e., $R_i \Leftrightarrow R_j$.

2. Rows may be mutiplied by a non-zero scalar, i.e., cR_i.

3. Rows and constant multiple of rows can be added or subtracted, i.e., $cR_i + R_j$.

To find the inverse of a matrix $\mathbf{A}$ of dimension n, we use elementary row operations to the matrix $[\mathbf{A}|\mathbf{I_n}]$ until we obtain $[\mathbf{I_n}|\mathbf{B}]$ where the matrix $\mathbf{B} = \mathbf{A}^{-1}$. Thus,

$$\begin{pmatrix} 3 & 7 & 10 & | & 1 & 0 & 0 \\ 7 & 17 & 25 & | & 0 & 1 & 0 \\ 10 & 25 & 38 & | & 0 & 0 & 1 \end{pmatrix}$$

can be transformed as follows:

1. $-\frac{7}{3}R_1 + R_2$ and $-\frac{10}{3}R_1 + R_3$

$$\begin{pmatrix} 3 & 7 & 10 & | & 1 & 0 & 0 \\ 0 & \frac{2}{3} & \frac{5}{3} & | & -\frac{7}{3} & 1 & 0 \\ 0 & \frac{5}{3} & \frac{14}{3} & | & -\frac{10}{3} & 0 & 1 \end{pmatrix}$$

2. $-\frac{21}{2}R_2 + R_1$ and $-\frac{5}{2}R_2 + R_3$

$$\begin{pmatrix} 3 & 0 & -\frac{15}{2} & | & \frac{51}{2} & -\frac{21}{2} & 0 \\ 0 & \frac{2}{3} & \frac{5}{3} & | & -\frac{7}{3} & 1 & 0 \\ 0 & 0 & \frac{1}{2} & | & \frac{5}{2} & -\frac{5}{2} & 1 \end{pmatrix}$$

3. $-15R_3 + R_1$ and $-\frac{10}{3}R_3 + R_2$

$$\begin{pmatrix} 3 & 0 & 0 & | & 63 & -48 & 15 \\ 0 & \frac{2}{3} & 0 & | & -\frac{32}{3} & \frac{28}{3} & -\frac{10}{3} \\ 0 & 0 & \frac{1}{2} & | & \frac{5}{2} & -\frac{5}{2} & 1 \end{pmatrix}$$

4. $\frac{1}{3}R_1$, $\frac{3}{2}R_2$ and $2R_3$

$$\begin{pmatrix} 1 & 0 & 0 & | & 21 & -16 & 5 \\ 0 & 1 & 0 & | & -16 & 14 & -5 \\ 0 & 0 & 1 & | & 5 & -5 & 2 \end{pmatrix}$$

(iv)

$$\mathbf{X}^T\mathbf{y} = \begin{pmatrix} 1 & 1 & 1 \\ 3 & 2 & 2 \\ 5 & 2 & 3 \end{pmatrix} \begin{pmatrix} 5 \\ 3 \\ 9 \end{pmatrix} = \begin{pmatrix} 17 \\ 39 \\ 58 \end{pmatrix}$$

$$\Rightarrow \mathbf{b} = \begin{pmatrix} 21 & -16 & 5 \\ -16 & 14 & -5 \\ 5 & -5 & 2 \end{pmatrix} \begin{pmatrix} 17 \\ 39 \\ 58 \end{pmatrix} = \begin{pmatrix} 23 \\ -16 \\ 6 \end{pmatrix}.$$

(v)

The first step is to determine whether or not $\mathbf{X}$ is *invertible*. If the determinant is non-zero then it is invertible. Expanding along the first row we have

$$\det(\mathbf{X}) = (1)(-1)^{1+1}\det\begin{pmatrix} 2 & 2 \\ 2 & 3 \end{pmatrix} + (3)(-1)^{1+2}\det\begin{pmatrix} 1 & 2 \\ 1 & 3 \end{pmatrix}$$

$$+(5)(-1)^{1+3}\det\begin{pmatrix} 1 & 2 \\ 1 & 2 \end{pmatrix}$$

$$\Rightarrow \quad \det(\mathbf{X}) = (1)(1)(2) + (3)(-1)(1) + (5)(1)(0) = -1.$$

Using the laws of matrix algebra for the determinant of the transpose we know that if $\mathbf{X}$ is invertible then so is $\mathbf{X^T}$. Using the rule that if two matrices $\mathbf{A}$ and $\mathbf{B}$ are invertible then $(\mathbf{AB})^{-1} = \mathbf{B}^{-1}\mathbf{A}^{-1}$.

$$\mathbf{b} = (\mathbf{X}^T\mathbf{X})^{-1}\mathbf{X}^T\mathbf{y} = \mathbf{X}^{-1}(\mathbf{X}^T)^{-1}\mathbf{X}^T\mathbf{y} = \mathbf{X}^{-1}\mathbf{y}.$$

Rank

Question 2

(i)

The rank of a matrix equals the number of linearly independent columns and is the order of the largest square matrix formed by deleting rows, columns or both that has a non-zero determinant.

Observation reveals that $c_2 = 2c_1$ and $c_3 = -4c_1$. Thus, there is only one independent column and the rank equals 1.

(ii)

For this matrix, there is no square matrix which can be formed by deleting rows, columns or both which has a non-zero determinant. Thus, the rank of the matrix is 0 (recall that the determinant of a square matrix of order one is the element itself).

(iii)

The *maximum* rank of a matrix is the minimum of {number of rows, number of columns}. Thus, the maximum rank of this matrix is 3. The matrix will have this rank if and only if we can form a (3×3) matrix which has a non-zero determinant by eliminating one of the columns of the original matrix.

If we eliminate the fourth column, the resulting (3×3) matrix is defined by $\mathbf{B}$

$$\mathbf{B} = \begin{pmatrix} 1 & 0 & 4 \\ 3 & 2 & 1 \\ 2 & 1 & 3 \end{pmatrix}.$$

Its determinant may be calculated by expanding along the first row, i.e.,

$$\det(\mathbf{B}) = 1(-1)^{1+1} \det \begin{pmatrix} 2 & 1 \\ 1 & 3 \end{pmatrix} + 0 + (4)(-1)^{1+3} \det \begin{pmatrix} 3 & 2 \\ 2 & 1 \end{pmatrix}$$

$$= 5 + 4(-1) = 1$$

Thus, the largest square matrix that we can form from the original matrix $\mathbf{A}$ by eliminating rows, columns or both which has a non-zero determinant is a (3×3) matrix. The rank of $\mathbf{A}$ is, therefore, $r(\mathbf{A}) = 3$.

Linear Independence and Rank

Question 3

(i)

Two vectors are *linearly independent* if the only way of making the sum $c_1\mathbf{v}_1 + c_2\mathbf{v}_2$ equal zero is by setting c_1 and c_2 both equal to zero. If the contrary is true, i.e., $c_1\mathbf{v}_1 + c_2\mathbf{v}_2 = \mathbf{0}$ for some $(c_1, c_2) \neq \mathbf{0}$, then it means that one vector is just a multiple of the other.

Thus the question is whether there exists $(c_1, c_2) \neq (0, 0)$ such that:

$$c_1 \begin{pmatrix} 5 \\ 1 \end{pmatrix} + c_2 \begin{pmatrix} -7 \\ 2 \end{pmatrix} = \begin{pmatrix} 0 \\ 0 \end{pmatrix}.$$

This condition can be restated as the following two equations:

$$5c_1 - 7c_2 = 0 \Rightarrow 5c_1 = 7c_2$$

$$c_1 + 2c_2 = 0 \Rightarrow c_1 = -2c_2$$

Substituting the second equation into the first, we obtain the condition

$$-10c_2 = 7c_2$$

This can only be true if $c_2 = 0$, otherwise we could divide both sides of the equation by c_2 and obtain the nonsense expression $-10 = 7$. However if $c_2 = 0$, then c_1 must equal zero also. Hence there is no $(c_1, c_2) \neq \mathbf{0}$ such that $c_1\mathbf{v}_1 + c_2\mathbf{v}_2 = \mathbf{0}$, and so $\mathbf{v}_1$ and $\mathbf{v}_2$ are linearly independent.

(ii)

Applying the same methodology as in the solution to (i) above, we look for two constants $(c_1, c_2) \neq \mathbf{0}$ that make true the following:

$$c_1 \begin{pmatrix} 5 \\ 1 \end{pmatrix} + c_2 \begin{pmatrix} -10 \\ -2 \end{pmatrix} = \begin{pmatrix} 0 \\ 0 \end{pmatrix}.$$

Restating the problem, we want to know when $5c_1 = 10c_2$ and $c_1 = 2c_2$ both hold true. In fact these equations hold for all c_1 and c_2, since they are effectively the same equation. Thus $\mathbf{v}_1$ and $\mathbf{v}_3$ are linearly dependent.

(iii)

The problem is to find the rank of the matrix

$$\mathbf{V} = [\mathbf{v}_1 : \mathbf{v}_2 : \mathbf{v}_3] = \begin{pmatrix} 5 & -7 & -10 \\ 1 & 2 & -2 \end{pmatrix}.$$

The *rank* of an n by m matrix is the maximum number of linearly independent columns. We know that $\mathbf{v}_1$ and $\mathbf{v}_2$ are linearly independent from (i), so the rank is at least 2. However we know from (ii) that $\mathbf{v}_1$ and $\mathbf{v}_3$ are linearly dependent, so the rank cannot be three. Thus the rank is two.

Solving Systems of Linear Equations

Question 4

(i)

Rewritten in the matrix form $\mathbf{Ax} = \mathbf{b}$ this system is equivalent to:

$$\begin{pmatrix} 1 & 1 & 0 & 1 \\ 1 & 0 & 1 & 1 \\ 2 & 1 & 1 & 2 \end{pmatrix} \begin{pmatrix} x_1 \\ x_2 \\ x_3 \\ x_4 \end{pmatrix} = \begin{pmatrix} 1 \\ 2 \\ 4 \end{pmatrix}.$$

For the system to have at least one solution we require that the rank of $\mathbf{A}$ ($rank(\mathbf{A})$) be equal to the rank of $[\mathbf{A}|\mathbf{b}]$ ($rank(\mathbf{A}|\mathbf{b})$) called the augmented coefficient matrix of the system. We can employ elementary row operations to find the rank by transforming the matrix $[\mathbf{A}|\mathbf{b}]$ to a *reduced* matrix.

Using elementary row operations on $[\mathbf{A}|\mathbf{b}]$, we obtain the follwoing:

1. $-R_1 + R_2$ and $-2R_2 + R_3$

$$\begin{pmatrix} 1 & 1 & 0 & 1 & | & 1 \\ 0 & -1 & 1 & 0 & | & 1 \\ 0 & -1 & 1 & 0 & | & 2 \end{pmatrix}$$

2. $-R_2 + R_3$ and $-R_2$

$$\begin{pmatrix} 1 & 1 & 0 & 1 & | & 1 \\ 0 & 1 & -1 & 0 & | & -1 \\ 0 & 0 & 0 & 0 & | & 1 \end{pmatrix}$$

The rank of the augmented coefficient matrix equals the number of non-zero rows in the *reduced* matrix above. There are three non-zero rows, thus, the rank is three. The rank of the coefficient matrix $[\mathbf{A}]$ is the number of non-zero rows to the left of the dotted line in the reduced matrix. There are two non-zero rows, thus, the rank of $[\mathbf{A}]$ is two.

A solution exists for a system of m linear equations and n unknowns *if and only if* rank$(\mathbf{A})$ = rank$(\mathbf{A}|\mathbf{b})$. Thus, there is no solution and the system of equations is *inconsistent*.

(ii)

Here the matrix $\mathbf{A}$ is given by

$$\mathbf{A} = \begin{pmatrix} 3 & 4 \\ 6 & 8 \end{pmatrix}$$

Because the second row equals the first row multiplied by two the rank of the matrix is one, which means that there is only one linearly dependent row in the matrix $\mathbf{A}$. What of the matrix $[\mathbf{A}|\mathbf{b}]$? We have

$$[\mathbf{A}|\mathbf{b}] = \begin{pmatrix} 3 & 4 & 4 \\ 6 & 8 & 8 \end{pmatrix}.$$

In this matrix also the second row is twice the first: hence the rank of $[\mathbf{A}|\mathbf{b}]$ is also one. Thus, rank $([\mathbf{A}|\mathbf{b}])$ = rank $(\mathbf{A})$ and there is at least one solution. However, since the rank of $\mathbf{A}$ is less than the dimension of $\mathbf{x}$ (note that dim$(\mathbf{x}) = 2$), the solution is not unique. Since $\mathbf{A}$ is square, i.e., the number of rows equals the number of columns, we could have come to the same conclusion by calculating the determinant of $\mathbf{A}$, which is equal to $(3)(8) - (6)(4) = 0$. A zero determinant implies that either the solution does not exist or the solution is **not** unique.

(iii)

Rewritten the system becomes

$$\mathbf{A}\mathbf{x} = \begin{pmatrix} 1 & 1 & 1 \\ -1 & 2 & -1 \\ -2 & 1 & -1 \end{pmatrix} \begin{pmatrix} x_1 \\ x_2 \\ x_3 \end{pmatrix} = \begin{pmatrix} 2 \\ 4 \\ 6 \end{pmatrix} = \mathbf{b}.$$

Because $\mathbf{A}$ is a square matrix we can apply the determinant test to see if $\mathbf{A}$ is singular. If $\mathbf{A}$ is singular then the determinant is zero and $\mathbf{A}^{-1}$ does not exist. In turn this implies that *no* solution unique exists to the system of linear equations. The determinant of $\mathbf{A}$ is a combination of the elements of row i of $\mathbf{A}, \mathbf{a}_{ij}$, and the cofactors of this row C_{ij}:

$$\det(\mathbf{A}) = \mathbf{a}_{i1}C_{i1} + \mathbf{a}_{i2}C_{i2} + \ldots + \mathbf{a}_{in}C_{in},$$

where

$$C_{ij} = (-1)^{i+j} \det(M_{ij}),$$

and M_{ij} is the matrix formed by deleting row i and column j from $\mathbf{A}$. Taking the first row since its elements are all equal to one (making multiplication easier), we have

$$\det(\mathbf{A}) = (1)(-1)^{1+1}\det\begin{pmatrix} 2 & -1 \\ 1 & -1 \end{pmatrix} + (1)(-1)^{1+2}\det\begin{pmatrix} -1 & -1 \\ -2 & -1 \end{pmatrix}$$
$$+(1)(-1)^{1+3}\det\begin{pmatrix} -1 & 2 \\ -2 & 1 \end{pmatrix}$$

It follows,

$$\det(\mathbf{A}) = (-2+1) + (-1)(1-2) + (-1+4)$$
$$\Rightarrow \det(\mathbf{A}) = -1+1+3 = 3.$$

Thus the matrix is non-singular and therefore we can invert $\mathbf{A}$ to give the solution to the system, $\mathbf{x} = \mathbf{A}^{-1}\mathbf{b}$.

The inverse matrix $\mathbf{A}^{-1}$ is calculated according to the formula

$$\mathbf{A}^{-1} = \frac{\text{adj}(\mathbf{A})}{\det(\mathbf{A})},$$

where the adjoint matrix $\text{adj}(\mathbf{A})$ is the transpose of the matrix of cofactors $[C_{ij}]$ whereby each row of $[C_{ij}]$ becomes the corresponding column of $\text{adj}(\mathbf{A})$.

Hence,

$$\mathbf{A}^{-1} = \frac{1}{3}\begin{pmatrix} -1 & 1 & 3 \\ 2 & 1 & -3 \\ -3 & 0 & 3 \end{pmatrix}^T = \begin{pmatrix} -1/3 & 2/3 & -1 \\ 1/3 & 1/3 & 0 \\ 1 & -1 & 1 \end{pmatrix}.$$

Thus the solution to the system of equations is

$$\mathbf{x} = \mathbf{A}^{-1}\mathbf{b} = \begin{pmatrix} -1/3 & 2/3 & -1 \\ 1/3 & 1/3 & 0 \\ 1 & -1 & 1 \end{pmatrix}\begin{pmatrix} 2 \\ 4 \\ 6 \end{pmatrix} = \begin{pmatrix} -4 \\ 2 \\ 4 \end{pmatrix}.$$

Input-Output Analysis

Question 5

(i)

Define the matrix of non-labor technical coefficients as

$$\mathbf{A} = \begin{pmatrix} 0.25 & 0.1 \\ 0.5 & 0.4 \end{pmatrix},$$

The Hawkins-Simon conditions require that the Leontief matrix $(\mathbf{I} - \mathbf{A})$ has the following characteristics:

1. All the elements on the main diagonal be positive.

2. All the leading principal minors be positive.

This requirement is true if and only if $1 - \mathbf{a}_{11} > 0$ and $1 - \mathbf{a}_{22} > 0$ and that $\det(1 - \mathbf{a}_{11}) > 0$ and $\det(\mathbf{I} - \mathbf{A}) > 0$. In our case, $0.75 > 0$ and $0.6 > 0$ and

$$
\begin{aligned}
\det(1 - \mathbf{a}_{11}) &= 0.75 > 0 \\
\det(\mathbf{I} - \mathbf{A}) &= \det\left(\begin{pmatrix} 1 & 0 \\ 0 & 1 \end{pmatrix} - \begin{pmatrix} 0.25 & 0.1 \\ 0.5 & 0.4 \end{pmatrix} \right) \\
&= \det\begin{pmatrix} 0.75 & -0.1 \\ -0.5 & 0.6 \end{pmatrix} \\
&= (0.75)(0.6) - (-0.1)(-0.5) \\
&= 0.45 - 0.05 = 0.4 > 0.
\end{aligned}
$$

Thus the conditions are satisfied.

(ii)

The first step is to determine the value of the non-labor input which will be required to satisfy the final demand vector $\mathbf{d}$. This is determined by the relation $(\mathbf{I} - \mathbf{A})\mathbf{x} = \mathbf{d}$ where $\mathbf{x}$ is a column vector of the total value of output in the two sectors, coal and steel. Using our knowledge of the inverse, the total value of the outputs is calculated as follows:

$$
\begin{aligned}
\mathbf{x} &= (\mathbf{I} - \mathbf{A})^{-1}\mathbf{d} \\
&= \frac{1}{0.4}\begin{pmatrix} 0.6 & 0.1 \\ 0.5 & 0.75 \end{pmatrix}\begin{pmatrix} 50 \\ 50 \end{pmatrix} \\
&= \begin{pmatrix} 87.5 \\ 156.25 \end{pmatrix}
\end{aligned}
$$

Because one dollar worth of coal requires 25 cents of labor input and one dollar worth of steel requires 50 cents of labor input, the total payment to labor is 100, i.e.,

$$
\begin{pmatrix} (87.5)(0.25) \\ (156.25)(0.50) \end{pmatrix} = \begin{pmatrix} 21.875 \\ 78.125 \end{pmatrix}.
$$

(iii)

The final consumption possibility set of the economy is the set of those possible final demands that do not exceed the total wage bill. In algebraic terms the set is defined below

$$
\{\mathbf{d} \geq \mathbf{0} : (\mathbf{I} - \mathbf{A})\mathbf{x} \geq \mathbf{d} \text{ and } \mathbf{a}_L\mathbf{x} \leq l\},
$$

where $\mathbf{a}_L$ is the labor requirement and W is the wage bill.

To determine if the final demand $\mathbf{d} = (20, 20)$ is feasible we must show that the total value of output multiplied by the value of the labor input in each sector is less than or equal to 100. The total value of output in the two sectors is calculated as follows:

$$
\mathbf{x} = (\mathbf{I} - \mathbf{A})^{-1}\mathbf{d}
$$

Feasibility requires that

$$\mathbf{a}_L\mathbf{x} = \mathbf{a}_L(\mathbf{I} - \mathbf{A})^{-1}\mathbf{d} \leq 100.$$

where $\mathbf{a}_L = (0.25, 0.50)$.

Substituting in the values for $\mathbf{A}$ and $\mathbf{a}_L$ we obtain

$$\left(\begin{array}{cc} 0.25 & 0.50 \end{array}\right) \frac{1}{0.4} \left(\begin{array}{cc} 0.6 & 0.1 \\ 0.5 & 0.75 \end{array}\right) \mathbf{d} \leq 100$$

$$\Rightarrow \left(\begin{array}{cc} 1.0 & 1.0 \end{array}\right) \mathbf{d} \leq 100.$$

If $\mathbf{d} = (20, 20)^T$, then

$$\left(\begin{array}{cc} 1.0 & 1.0 \end{array}\right) \mathbf{d} = \left(\begin{array}{cc} 1.0 & 1.0 \end{array}\right) \left(\begin{array}{c} 20 \\ 20 \end{array}\right) = 20.0 + 20.0 = 40.0 < 100.$$

Thus, the final demand vector (20,20) is feasible.

National Income Model

Question 6

The first step is to rewrite the system in the form $\mathbf{A}\mathbf{x} = \mathbf{b}$, where $\mathbf{A}$ is the matrix of coefficients in the system of equations, $\mathbf{x}$ is the vector of endogenous variables which must be solved for, and $\mathbf{b}$ is the vector of constants which includes the exogenous variables and parameters. In forming $\mathbf{A}$, the first column corresponds to the coefficients on Y in the system of equations, the second column corresponds to the coefficients on C, the third to I and the fourth to R.

$$\left(\begin{array}{cccc} 1 & -1 & -1 & 0 \\ -\beta(1-\tau) & 1 & 0 & \gamma \\ -\epsilon(1-\tau) & 0 & 1 & \zeta \\ \theta(1-\tau) & 0 & 0 & -\eta \end{array}\right) \left(\begin{array}{c} Y \\ C \\ I \\ R \end{array}\right) = \left(\begin{array}{c} G \\ \alpha \\ \delta \\ M \end{array}\right)$$

In order to answer the question it is not necessary to invert the entire matrix $\mathbf{A}$. This is because we are not interested in finding the solutions of **all** the endogenous variables, but just the solution for Y. To do this we can employ *Cramer's rule*, which states that if $\mathbf{x} = \mathbf{A}^{-1}\mathbf{b}$, then the ith element of $\mathbf{x}$ is given by

$$\mathbf{x}_i = \frac{\det(\mathbf{B}_i)}{\det(\mathbf{A})},$$

The matrix $\mathbf{B}_i$ is the matrix $\mathbf{A}$ with the ith column replaced by the vector $\mathbf{b}$. In our case,

$$\mathbf{B}_1 = \left(\begin{array}{cccc} G & -1 & -1 & 0 \\ \alpha & 1 & 0 & \gamma \\ \delta & 0 & 1 & \zeta \\ M & 0 & 0 & -\eta \end{array}\right).$$

To find the determinant of $\mathbf{B}_1$ we expand along the bottom row of the matrix:

$$\det(\mathbf{B}_1) = M \det \begin{pmatrix} -1 & -1 & 0 \\ 1 & 0 & \gamma \\ 0 & 1 & \zeta \end{pmatrix} - \eta \det \begin{pmatrix} G & -1 & -1 \\ \alpha & 1 & 0 \\ \delta & 0 & 1 \end{pmatrix}$$

$$= +M(\gamma + \zeta) - (-\eta)(\alpha + \delta + G)$$

The next step is to evaluate the determinant of the matrix $\mathbf{A}$. Once more we expand along the bottom row:

$$\det(\mathbf{A}) = \theta(1 - \tau) \det \begin{pmatrix} -1 & -1 & 0 \\ 1 & 0 & \gamma \\ 0 & 1 & \zeta \end{pmatrix} - (-\eta) \det \begin{pmatrix} 1 & -1 & -1 \\ -\beta(1-\tau) & 1 & 0 \\ -\epsilon(1-\tau) & 0 & 1 \end{pmatrix}$$

$$= \theta(1 - \tau)(\gamma + \zeta) + \eta(-\beta(1 - \tau) - \epsilon(1 - \tau) + 1)$$

Assuming that the expression above is not zero, we can now form the expression for Y:

$$Y = \frac{M(\gamma + \zeta) + (\alpha + \delta + G)\eta}{\eta + (1 - \tau)[\theta(\gamma + \zeta) - \eta(\beta + \epsilon)]}.$$

This formula is known as the *reduced form* of Y, since it gives the value of Y in terms of the exogenous variables and the parameters only—given these values we are able to solve for the numerical value of Y.

Eigenvalues

Question 7

(i)

$\mathbf{A}_1$: We need to find the values of λ that lead to the equality $\mathbf{A}_1 \mathbf{x} = \lambda \mathbf{x}$, for some $\mathbf{x}$. This is equivalent to solving a homogeneous system of linear equations, i.e., the values of λ that ensure a non-trivial solution exists for $\mathbf{x}$. These values will be the eigenvalues of the system.

Rewritten, the equation is $(\mathbf{A}_1 - \lambda \mathbf{I})\mathbf{x} = \mathbf{0}$. For this to hold it must be that the matrix $(\mathbf{A}_1 - \lambda \mathbf{I})$ is singular, and thus $\det(\mathbf{A}_1 - \lambda \mathbf{I}) = 0$. Using this is the fact we can compute the eigenvalues.

$$\det(\mathbf{A}_1 - \lambda \mathbf{I}) = \det \begin{pmatrix} 4 - \lambda & -5 \\ 2 & -3 - \lambda \end{pmatrix}$$

$$\Rightarrow \det(\mathbf{A}_1 - \lambda \mathbf{I}) = (4 - \lambda)(-3 - \lambda) - (2)(-5)$$

$$\Rightarrow \det(\mathbf{A}_1 - \lambda \mathbf{I}) = \lambda^2 - \lambda - 2.$$

To solve the equation we may use the quadratic formula or factorize to solve for the characteristic equation to obtain the solutions. Using the quadratic formula:

$$\lambda = \frac{1 \pm \sqrt{1^2 - 4(-2)}}{2} = 2 \text{ or } \text{-1}.$$

These two numbers are the eigenvalues of $\mathbf{A}_1$.

$\mathbf{A}_2$: Using the same method as above, we have

$$\det(\mathbf{A}_2 - \lambda\mathbf{I}) = \det\begin{pmatrix} 1-\lambda & 2 \\ -1 & 3-\lambda \end{pmatrix}$$
$$\Rightarrow \det(\mathbf{A}_2 - \lambda\mathbf{I}) = (1-\lambda)(3-\lambda) - (2)(-1)$$
$$\Rightarrow \det(\mathbf{A}_2 - \lambda\mathbf{I}) = \lambda^2 - 4\lambda + 5.$$

Solving this equation using the quadratic formula:

$$\lambda = \frac{4 \pm \sqrt{16-20}}{2} = 2 \pm i.$$

Thus the eigenvalues associated with $\mathbf{A}_2$ are complex numbers, where $i = \sqrt{-1}$ (see chapter nine).

$\mathbf{A}_3$: Calculating the determinant in this case is not difficult because the second row of $(\mathbf{A}_3 - \lambda\mathbf{I})$ has only one non-zero value. Thus:

$$\det(\mathbf{A}_3 - \lambda\mathbf{I}) = \det\begin{pmatrix} 1-\lambda & 0 & 2 \\ 0 & 1-\lambda & 0 \\ 2 & 0 & 1-\lambda \end{pmatrix}$$
$$= 0 + (1-\lambda)(-1)^{2+2}\det\begin{pmatrix} 1-\lambda & 2 \\ 2 & 1-\lambda \end{pmatrix} + 0$$
$$= (1-\lambda)[(1-\lambda)^2 - 4].$$

One root of this equation is $\lambda = 1$. The other roots can be found by setting $(1-\lambda)^2 - 4$ to zero and solving:

$$(1-\lambda)^2 - 4 = 0$$
$$\Rightarrow (1-\lambda)^2 = 4$$
$$\Rightarrow (1-\lambda) = \pm 2$$
$$\Rightarrow \lambda = -1 \text{ or } 3.$$

(ii)

$\mathbf{A}_1$: Recall that for equations of the form

$$\mathbf{Ax} = \lambda\mathbf{x}$$

λ and $\mathbf{x}$ are known, repectively, as the eigenvalues and eigenvectors of the system. From (i) we know that the possible eigenvalues for $\mathbf{A}_1$ are $\lambda_1 = -1$ and $\lambda_2 = 2$. The task now is to find the eigenvectors.

Starting with $\lambda_1 = -1$,

$$\mathbf{A}_1\mathbf{x} = \lambda_1\mathbf{x}$$
$$\Rightarrow (\mathbf{A}_1 - \lambda_1\mathbf{I})\mathbf{x} = \mathbf{0}$$
$$\Rightarrow \begin{pmatrix} 4+1 & -5 \\ 2 & -3+1 \end{pmatrix}\mathbf{x} = \mathbf{0}$$

This implies that $5x_1-5x_2 = 0$ and $2x_1-2x_2 = 0$. These two equations both imply that $x_1 = x_2$. This is not an accident. The whole point of an eigenvector is to make the rows of $\mathbf{A} - \lambda\mathbf{I}$ linearly dependent, and thus to make the matrix $\mathbf{A} - \lambda\mathbf{I}$ singular. The set of eigenvectors corresponding to λ_1 is therefore shown to be all vectors $\mathbf{x}$ in the set of two-dimensional real numbers such that $x_1 = x_2$.

Using now the second eigenvalue,

$$(\mathbf{A}_1 - \lambda_2\mathbf{I})\mathbf{x} = \mathbf{0}$$

$$\Rightarrow \begin{pmatrix} 4-2 & -5 \\ 2 & -3-2 \end{pmatrix} \mathbf{x} = \mathbf{0}$$

From the above we obtain the following equation $2x_1 - 5x_2 = 0$. Thus, the set of eigenvectors corresponding to λ_2 is all $\mathbf{x}$ in the set of two-dimensional real numbers such that $x_1 = 5x_2/2$.

$\mathbf{A}_2$: The eigenvalues in this case contain an imaginary number and are therefore not real numbers.

$\mathbf{A}_3$: In this case we found the possible eigenvalues are $\{1, -1, 3\}$. For $\lambda_1 = 1$:

$$(\mathbf{A}_3 - \lambda_1\mathbf{I})\mathbf{x} = \mathbf{0}$$

$$\Rightarrow \begin{pmatrix} 1-1 & 0 & 2 \\ 0 & 1-1 & 0 \\ 2 & 0 & 1-1 \end{pmatrix} \mathbf{x} = \mathbf{0}$$

$$\Rightarrow 2x_1 = 0 \text{ and } 2x_3 = 0.$$

Thus the eigenvectors associated with λ_1 are all the vectors in the set of three-dimensional real numbers of the form $(0, \alpha, 0)$, where α can be any real number. Setting $\lambda_2 = -1$ implies that

$$(\mathbf{A}_3 - \lambda_2\mathbf{I})\mathbf{x} = \mathbf{0}$$

$$\Rightarrow \begin{pmatrix} 1+1 & 0 & 2 \\ 0 & 1+1 & 0 \\ 2 & 0 & 1+1 \end{pmatrix} \mathbf{x} = \mathbf{0}$$

which implies

$$2x_1 = -2x_3$$
$$x_2 = 0$$
$$2x_1 = -2x_3$$

The eigenvectors are therefore of the form $\mathbf{x} = (\alpha, 0, -\alpha)$. Finally, with $\lambda_3 = 3$ we have

$$(\mathbf{A}_3 - \lambda_3\mathbf{I})\mathbf{x} = \mathbf{0}$$

$$\Rightarrow \begin{pmatrix} 1-3 & 0 & 2 \\ 0 & 1-3 & 0 \\ 2 & 0 & 1-3 \end{pmatrix} \mathbf{x} = \mathbf{0}$$

which implies

$$2x_1 = 2x_3$$
$$x_2 = 0$$
$$2x_1 = 2x_3$$

This gives eigenvectors of the form $\mathbf{x} = (\alpha, 0, \alpha)$, where α is any real number.

(iii)

$\mathbf{A}_1$: *A* set of eigenvectors is *orthonormal* if each vector is of unit length and if each vector is orthogonal to the others. In the case of $\mathbf{A}_1$, the set of eigenvectors is given by

$$\mathbf{x}^1 = \begin{pmatrix} \alpha \\ \alpha \end{pmatrix}, \mathbf{x}^2 = \begin{pmatrix} \beta \\ 2\beta/5 \end{pmatrix}, \text{ where } \alpha, \beta \in \text{ set of real numbers.}$$

Note that these expressions define a *family* of vectors. In the case of $\mathbf{x}^1$, it is all vectors from the origin with a slope of +1. Although the lengths are different, the direction is the same, and direction is what matters for orthogonality. According to the definition, two vectors $\mathbf{x}^1$ and $\mathbf{x}^2$ are *orthogonal* if the inner product $\mathbf{x}^{1^T}\mathbf{x}^2 = 0$. In two dimensions this definition requires that the vectors be at right angles, or *perpendicular*, to each other. In our case the inner product is

$$\mathbf{x}^{1^T}\mathbf{x}^2 = \begin{pmatrix} \alpha & \alpha \end{pmatrix} \begin{pmatrix} \beta \\ 2\beta/5 \end{pmatrix} = \alpha\beta + \frac{2}{5}\alpha\beta$$

$$\Rightarrow \mathbf{x}^{1^T}\mathbf{x}^2 = \frac{3}{5}\alpha\beta.$$

This expression can only be zero if either α or β equals zero. However in that case the length of one of the vectors would be zero: thus none of the eigenvectors are orthonormal, since for that property to hold we require that *both* vectors be of unit length.

$\mathbf{A}_3$: Taking now the eigenvectors associated with the matrix $\mathbf{A}_3$,

$$\mathbf{x}^1 = \begin{pmatrix} 0 \\ \alpha \\ 0 \end{pmatrix}, \mathbf{x}^2 = \begin{pmatrix} \beta \\ 0 \\ -\beta \end{pmatrix}, \mathbf{x}^3 = \begin{pmatrix} \gamma \\ 0 \\ \gamma \end{pmatrix}, \alpha, \beta, \gamma \in \text{ set of real numbers.}$$

First we check to see when the families of vectors are orthogonal i.e., when are the inner products equal to zero?

$$\mathbf{x}^{1^T}\mathbf{x}^2 = 0 \cdot \beta + \alpha \cdot 0 + 0 \cdot \beta = 0,$$
$$\mathbf{x}^{2^T}\mathbf{x}^3 = \beta \cdot \gamma + 0 \cdot 0 - \beta \cdot \gamma = 0,$$
$$\mathbf{x}^{3^T}\mathbf{x}^1 = \gamma \cdot 0 + 0 \cdot \alpha - \gamma \cdot 0 = 0.$$

Thus the families of vectors are orthogonal for all $\alpha, \beta, \gamma \in$ set of real numbers. For what values of α, β and γ do the vectors have unit length?

$$|\mathbf{x}^1| = \sqrt{0^2 + \alpha^2 + 0^2} = \alpha$$
$$|\mathbf{x}^2| = \sqrt{\beta^2 + 0^2 + \beta^2} = \beta\sqrt{2}$$
$$|\mathbf{x}^3| = \sqrt{\gamma^2 + 0^2 + \gamma^2} = \gamma\sqrt{2}.$$

To normalize these vectors to be of unit length it is necessary to set $\alpha = 1, \beta = \sqrt{2}$, and $\gamma = \sqrt{2}$. We can now conclude that the following set of eigenvectors of $\mathbf{A}_3$ are orthonormal:

$$\mathbf{x}^1 = \begin{pmatrix} 0 \\ 1 \\ 0 \end{pmatrix},$$

$$\mathbf{x}^2 = \begin{pmatrix} 1/\sqrt{2} \\ 0 \\ -1/\sqrt{2} \end{pmatrix},$$

$$\mathbf{x}^3 = \begin{pmatrix} 1/\sqrt{2} \\ 0 \\ 1/\sqrt{2} \end{pmatrix}.$$

Quadratic Forms

Question 8

(i)

We want to convert the equation $f(\mathbf{x})$ into the quadratic form $Q(x) = \mathbf{x}^T \mathbf{A}\mathbf{x}$. For the two variable cases the quadratic form is

$$ax_1^2 + 2bx_1x_2 + cx_2^2 = \begin{pmatrix} x_1 & x_2 \end{pmatrix} \begin{pmatrix} a & b \\ b & c \end{pmatrix} \begin{pmatrix} x_1 \\ x_2 \end{pmatrix}.$$

For this problem, $a = 1, b = 1$ and $c = 1$. Thus, matrix $\mathbf{A}$ is defined as

$$\begin{pmatrix} 1 & 1 \\ 1 & 1 \end{pmatrix}$$

and the quadratic form of $f(\mathbf{x})$ is

$$\mathbf{x}^T \mathbf{A}\mathbf{x} = \begin{pmatrix} x_1 & x_2 \end{pmatrix} \begin{pmatrix} 1 & 1 \\ 1 & 1 \end{pmatrix} \begin{pmatrix} x_1 \\ x_2 \end{pmatrix}.$$

To test the definiteness of $\mathbf{A}$, we will calculate the leading principal minors of $\mathbf{A}$.

$$\det(\mathbf{a}_{11}) = \mathbf{a}_{11} = 1,$$
$$\det(\mathbf{A}) = 1 - 1 = 0.$$

The matrix fails the test for positive definiteness as the second leading principal minor is 0. To test for positive semidefiniteness we must examine **all** the principal minors and ensure that they are ≥ 0. In the case of a two by two matrix we only need to verify that each element of the main diagonal and the determinant of the matrix are ≥ 0. Thus, the matrix is positive semidefinite as is the quadratic form.

(ii) The quadratic form of $f(\mathbf{x})$ is

$$\mathbf{x}^T \mathbf{A} \mathbf{x} = \begin{pmatrix} x_1 & x_2 \end{pmatrix} \begin{pmatrix} 1 & -1 \\ -1 & 1 \end{pmatrix} \begin{pmatrix} x_1 \\ x_2 \end{pmatrix}.$$

The successive leading principal minors of the matrix $\mathbf{A}$ are

$$\det(\mathbf{a}_{11}) = 1,$$
$$\det(\mathbf{A}) = 1 - 1 = 0.$$

The matrix fails the test for positive definiteness as the second leading principal minor is 0. To test for positive semidefiniteness we must examine *all* the principal minors and ensure that they are ≥ 0. In the case of a two by two matrix we only need to verify that each element of the main diagonal and the determinant of the matrix are ≥ 0. Thus, the matrix $\mathbf{A}$ is positive semidefinite and so is the quadratic form.

(iii) The quadratic form of $f(\mathbf{x})$ is

$$\mathbf{x}^T \mathbf{A} \mathbf{x} = \begin{pmatrix} x_1 & x_2 \end{pmatrix} \begin{pmatrix} -2 & -1 \\ -1 & -1 \end{pmatrix} \begin{pmatrix} x_1 \\ x_2 \end{pmatrix}.$$

The successive leading principal minors of the matrix $\mathbf{A}$ are

$$\det(\mathbf{a}_{11}) = -2,$$
$$\det(\mathbf{A}) = 2 - 1 = 1.$$

Thus the matrix is negative definite.

(iv) In this case the quadratic form is

$$\mathbf{x}^T \mathbf{A} \mathbf{x} = \begin{pmatrix} x_1 & x_2 & x_3 \end{pmatrix} \begin{pmatrix} 2 & 1 & 0 \\ 1 & 1 & -1/2 \\ 0 & -1/2 & 2 \end{pmatrix} \begin{pmatrix} x_1 \\ x_2 \\ x_3 \end{pmatrix}$$

The successive leading principal minors are as follows:

$$\det(\mathbf{a}_{11}) = 2,$$
$$\det \begin{pmatrix} \mathbf{a}_{11} & \mathbf{a}_{12} \\ \mathbf{a}_{21} & \mathbf{a}_{22} \end{pmatrix} = \det \begin{pmatrix} 2 & 1 \\ 1 & 1 \end{pmatrix} = 2 - 1 = 1$$
$$\det(\mathbf{A}) = 2(-1)^{1+1}(2 - 1/4) + 1(-1)^{1+2}(2 - 0) = 3.5 - 2 = 1.5.$$

Thus $\mathbf{A}$ and the quadratic form are positive definite because all of the leading principal minors of $\mathbf{A}$ are positive.

Chapter 3

Functions and Sets

Objectives

The questions in this chapter should help the reader understand the fundamentals of set theory and those properties of functions, such as continuity and differentiability, that are important in economic applications. Readers who can correctly answer all the questions should be able to:

1. Define the basic properties of sets such as openness, boundedness and convexity (Question 1).

2. Identify increasing functions (Question 10).

3. Evaluate limits (Question 5).

4. Use the natural exponential function (Questions 2 and 3)

5. Apply logarithms to determine elasticities(Question 4)

6. Apply the concept of continuity (Question 7).

7. Apply the concept of differentiability (Questions 8 and 9).

8. Identify the curvature of functions (Question 11).

9. Determine the homogeneity and homotheticity of functions (Questions 12 and 13).

10. Apply Euler's Theorem (Question 14)

11. Apply the Intermediate Value Theorem (Question 6).

12. Use the Implicit Function Theorem (Questions 15 and 16).

13. Apply Taylor's theorem (Question 17).

Review

Functions and sets are the basic building blocks of mathematics, and thus of mathematical economics. *Sets* are collections of objects: an example might be the set of commodities consumed by an agent. If an object x is a member of a set S, then we use the following notation:

$$x \in S,$$

where the symbol $\in$ means *is an element of*. Other important symbols in set theory include: $\mathbf{A} \cup \mathbf{B}$ which denotes the set containing **all** the elements that belong to at least one of the sets $\mathbf{A}$ and $\mathbf{B}$; $\mathbf{A} \cap \mathbf{B}$ which is the set of elements that belong to **both** $\mathbf{A}$ and $\mathbf{B}$; and $\mathbf{A} \setminus \mathbf{B}$ which is the set of elements that belong to $\mathbf{A}$ but not to $\mathbf{B}$.

A set can be defined merely by listing its components, e.g.

$$S = \{a, b, c, d\}.$$

Sometimes, however, a set has too many elements to be easily defined in such a manner, and in this case we specify a set by the properties that its members must have. For example, the set of all strictly positive real numbers is defined as:

$$\mathcal{R}_{++} = \{x \in \mathcal{R} : x > 0\}.$$

where, in general, we define a set by { typical member: defining properties of the set }. In words, the set $\mathcal{R}_{++}$ consists of all real numbers ($x \in \mathcal{R}$) that are strictly positive ($x > 0$) where the set of all real numbers includes all *integers*, *rational* and *irrational* numbers. The set of integers includes $\cdots - 2, -1, 0, 1, 2 \cdots$; the set of rational numbers are all numbers formed by any ratio of integers defined as $\frac{a}{b}$ where $b \neq 0$ and the set of irrational numbers are those real numbers that cannot be formed as a ratio of integers. Examples are π, e and $\sqrt{2}$.

Another frequently used set is the set of all two-dimensional numbers $\mathcal{R}^2$. Elements of this set take the form of Cartesian coordinates, for example $(x_1, x_2) = (-1, 3)$ or $(0, 1000)$, and represent points in two dimensional real space. We often represent coordinates in vector form: $(x_1, x_2) = \mathbf{x}$.

Sets may have various properties. An *open* set is one that does not contain its boundary. A point in the boundary of a set S has the characteristic that if we draw a circle, no matter how small but with a radius $\epsilon > 0$, the circle will contain at least one point within the set and one point outside of the set. For example:

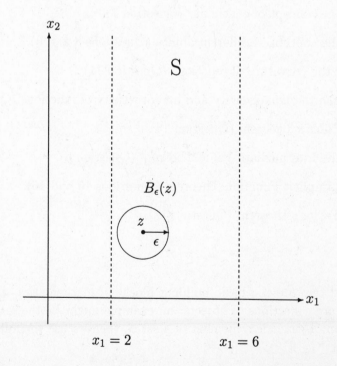

This set is defined by:

$$S = \{(x_1, x_2) \in \mathcal{R}^2 : 2 < x_1 < 6\},$$

The boundary of S is the two lines $x_1 = 2$ and $x_1 = 6$. To see that S is indeed open, choose any point in this set, the point z illustrated in the diagram, and draw a ball of radius $\epsilon > 0$ around it. This ball is labelled $B_\epsilon(z)$. If the set is open, then as long as epsilon is small enough, the whole ball should be within the set.

The opposite of an open set is a *closed* set, defined as a set that contains its boundary. The set S is not closed since it does not contain its boundaries $x_1 = 2$ and $x_1 = 6$. An example of a set that is closed is

$$\{x \in \mathcal{R} : 0 \le x \le 1\}$$

This is the set of all real numbers between 0 and 1, **including** 0 and 1. This set is often written as [0,1]. Similarly the open set

$$\{x \in \mathcal{R} : 0 < x < 1\}$$

is often written as (0,1). Note that a set may be neither open nor closed. For example the set [0,1), defined as

$$\{x \in \mathcal{R} : 0 \le x < 1\},$$

is not closed since it does not contain the boundary point $x = 1$. However neither is it open, since it *does* contain the boundary point $x = 0$.

A set is *convex* if for any two points $\mathbf{x}$ and $\mathbf{y}$ in the set,

$$\mathbf{z}(\lambda) = \lambda \cdot \mathbf{x} + (1 - \lambda) \cdot \mathbf{y} \in S$$

for all $\lambda \in [0,1]$. Intuitively, this means that if we draw a line between any two points in the set, the line should remain within the set at all points.

A set is *bounded* if the whole set can be contained within a sufficiently large enough circle. Thus the set

$$\{x \in \mathcal{R} : 0 \le x < 1\},$$

is bounded because it can be enclosed in a circle. However, the set

$$\{x \in \mathcal{R} : x > 0\},$$

is not bounded as there is no circle large enough to enclose the set. When determining whether a set is bounded or not we can also use the *distance formula* (d) which defines the distance between two points (x_1, y_1) and (x_2, y_2) as

$$d = \sqrt{(x_1 - x_2)^2 + (y_1 - y_2)^2}$$

A set is bounded if the distance betwen any pair of points in the set is not greater than some finite number, r. A set that is both bounded and closed is sometimes known as a *compact* set.

A *function* is a relation between at least two variables. For the function defined as $y = f(x)$ the variable x is the *domain* of the function or the set of all permissible values of x and the variable y is the *range* of the function or all the possible values into which x may be transformed or mapped.

A function always has a *unique* value of y for each x such that there cannot be more than one y value for every x. There may, however, be more than one value of x that gives the same y value. A function $f(x)$ which only has one value of x for every y is said to have an *inverse* function defined as $f^{-1}(x)$. An example of a function from macroeconomics is the consumption function.

$$C = f(Y^D)$$

which relates individuals' disposable income (Y^D) to their total consumption (C).

If a function has certain properties then it can make economic analysis much easier. For example, if the consumption function is *strictly increasing* then we know that an increase in income will raise consumption. Similarly, if the function is *continuous* then we know that the increase in consumption will be small if the change in income was small enough. Knowing the *derivative* enables us to derive a precise value for the change.

A function may be either *increasing*, *strictly increasing*, *decreasing* or *strictly decreasing*. For a function $f(x)$ and where $a < b$:

1. If $f(a) \leq f(b)$ then $f(x)$ is increasing

2. If $f(a) < f(b)$ then $f(x)$ is strictly increasing

3. If $f(a) \geq f(b)$ then $f(x)$ is decreasing

4. If $f(a) > f(b)$ then $f(x)$ is strictly decreasing

Functions that are strictly increasing are sometimes described as *monotonically* increasing and functions which are strictly decreasing are described as monotonically decreasing. If a function $f(x)$ is a continuous and differentiable in the interval [a,b] then if

1. $f'(x) > 0$ for all x in the interval (a,b), $f(x)$ is strictly increasing.

2. $f'(x) < 0$ for all x in the interval (a,b), $f(x)$ is strictly decreasing.

The first derivative conditions are *sufficient* in that if they hold true the function is either strictly increasing or decreasing. The conditions, however, are not *necessary* in that a function may still be strictly increasing or decreasing and not satisfy the conditions.

To understand the concepts of continuity and differentiability we must first understand the concept of the *limit* of a sequence. Consider the sequence

$$1, 1/2, 1/4, 1/8, 1/16, \ldots\ldots$$

The limit of this sequence of fractions is 0. By this we mean that the sequence is *converging* to 0. If we construct an interval of radius ϵ around 0, no matter how small we make ϵ, the sequence will eventually enter this interval, as in the diagram below:

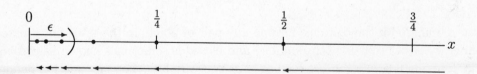

Note that the sequence will never actually get to zero: there will always be some distance, however infinitesimally small, between the sequence and zero. However, that distance can be made as small as we like, and this is what is required in the definition of a limit.

A related concept is the limit of a function. The limit of $f(x)$ as a sequence x goes to a point z is denoted by:

$$\lim_{x \to z} f(x)$$

For a limit to exist it must be true that the left hand and right hand limits are identical. The *left hand limit* is written as

$$\lim_{x \to z^-} f(x).$$

The minus sign indicates we are considering what happens to $f(x)$ as we approach the number in question (z) from a lower value. The *right hand limit* is what happens to $f(x)$ as we approach z from a higher value.

$$\lim_{x \to z^+} f(x).$$

If $\lim_{x \to z^-} f(x) = \lim_{x \to z^+} f(x) = \lim_{x \to z} f(x) = A$, then a *single* limit of the function exists at z if A is a finite number. A function is *continuous* at a point z if

1. z is in the *domain* of the function and, hence, is defined at z.

2. $\lim_{x \to z} f(x)$ exists.

3. $\lim_{x \to z} f(x) = f(z)$.

Graphically, a continuous function can be drawn without removing one's pen from the paper. A useful rule for continuous functions is that a composite function of two continuous functions defined by $f(x)$ and $g(x)$, such that $h(x) = f(g(x))$, is continuous.

Because a sequence $x_1, x_2, x_3, \ldots$ is converging to z, it does *not* necessarily follow that the limit of $f(x_1), f(x_2), f(x_3), \ldots$ is $f(z)$. This will only be true if the function is *continuous*. Consider the discontinuous function

$$f(x) = \begin{cases} 1, x > 0 \\ 0, x \leq 0 \end{cases}$$

and the sequence we examined previously, $1, \frac{1}{2}, \frac{1}{4}, \frac{1}{8}, \ldots$ In this case, the limit of the sequence $f(1) = 1, f(\frac{1}{2}) = 1, f(\frac{1}{4}) = 1, f(\frac{1}{8}) = 1$, is 1; not $f(0) = 0$.

Some useful rules when solving the limits of a function are listed below where K is a constant and $\lim_{x \to z} f(x) = A$ and $\lim_{x \to z} g(x) = B$.

1. $\lim_{x \to z} K f(x) = KA$.

2. $\lim_{x \to z} [f(x) \pm g(x)] = A \pm B$.

3. $\lim_{x \to z} f(x) g(x) = AB$.

4. $\lim_{x \to z} \frac{f(x)}{g(x)} = \frac{A}{B}$.

5. $\lim_{x \to z} [f(x)]^j = A^j$.

6. If $f(x)$ and $g(x)$ are equal for all values of x that are close to z, but not necessarily at z, then $\lim_{x \to z} f(x) = \lim_{x \to z} g(x)$ whenever the limit exists.

Another useful rule when evaluating the limit of the quotient of two functions is L'Hôpital's rule. It is used, in particular, when the numerator and denominator are not finite valued or when the numerator or denominator equals 0.

$$\lim_{x \to a} \frac{f(x)}{g(x)} = \lim_{x \to a} \frac{f'(x)}{g'(x)}$$

where $g'(x) \neq 0$ for all $x \neq a$ and $f(x)$ and $g(x)$ are differentiable in an interval around a but not necessarily at a.

In the case where $f(x)$ and $g(x)$ are differentiable at a and $f(a) = g(a) = 0$ then provided that $g'(a) \neq 0$

$$\lim_{x \to a} \frac{f(x)}{g(x)} = \lim_{x \to a} \frac{f'(a)}{g'(a)}$$

The *derivative* of a function is the rate of change of the function with respect to its argument. The formal definition is defined below in terms of the *difference quotient*.

$$\frac{df(x)}{dx} = f'(x) = \lim_{\Delta x \to 0} \frac{f(z + \Delta x) - f(z)}{\Delta x}$$

where $\frac{df(x)}{dx}$ exists at $x = z$ if and only if the limit exists as $\Delta x \to 0$.

A function is *differentiable* at a point z if it is continuous and if the slope of the function to the left of z is equal to the slope of the function to the right of z, where z can be any point. To check if a function is differentiable, first, verify if it is continuous and, second, determine if the left and right side limits of the difference quotient are identical and finite valued. Calculating the left and right side limits is easier if we redefine the difference quotient for $f(x)$ at $x = z$ as follows:

$$\lim_{x \to z} \frac{f(x) - f(z)}{x - z}$$

Thus, the difference quotient of the function $y = x^2$ at $x = 2$ is

$$\lim_{x \to 2} \frac{f(x) - f(2)}{x - 2}$$

where for $x < 2$ such that $x = \{1, 1.5.1.9, 1.99\}$ we have the following values for the difference quotient $\{3, 3.5, 3.9, 3.99\}$ and where $x > 2$ such that $x = \{3, 2.5, 2.1, 2.01\}$ we have the following values for the difference quotient $\{5, 4.5, 4.1, 4.01\}$. Thus,

$$\lim_{x \to 2^-} \frac{f(x) - f(2)}{x - 2} = \lim_{x \to 2^+} \frac{f(x) - f(2)}{x - 2} = 4$$

Because $y = x^2$ is continuous and the left and right hand limits of the difference quotient are identical and finite valued at $x = 2$, then the function is differentiable at this point.

The *stationary point* of a differentiable function $f(x)$ is the value of x where $f'(x) = 0$. To obtain the derivatives of functions we use a number of rules. One of the most useful rules is the derivative of a variable raised to a power, the *power rule*:

$$\frac{dx^n}{dx} = nx^{n-1}.$$

Also useful are the rules for taking the derivative of the natural logarithmic and exponential functions:

$$\frac{d\ln(x)}{dx} = \frac{1}{x},$$
$$\frac{de^x}{dx} = e^x.$$

In general, an *exponential* function is written as

$$f(x) = Ab^x$$

where A is a constant, b is the base of the function and x is the *exponent*. In the case where the base equals $e = 2.71828\ldots$ we obtain the natural exponential base. Formally we may define the irrational number e, sometimes written as exp, as follows:

$$e \equiv \lim_{n \to \infty} (1 + \frac{1}{n})^n$$

The number e has a convenient economic interpretation. Suppose we have a principal of \$1 which we wish to invest for a year at the interest rate of 100%. How much money we will receive at the end of the year depends upon how often the interest is compounded. If interest is only calculated at the end of the year, we will receive

$$\$1(1 + 100\%/1) = \$2$$

If interest is compounded twice a year, then we will receive

$$\$1(1 + 100\%/2)(1 + 100\%/2) = \$1(1.5^2) = \$2.25$$

In general then, if the interest is compounded n times a year, we will receive

$$\$1(1 + 100\%/n)^n$$

Thus, if the interest is compounded infinitely often (in other words, continuously), we will receive \$e.

We can use the number e to calculate continuously compounded interest in more general cases. If we have a principal of A that we wish to invest for n years at an interest rate of r, then the value V of the investment at the end of the n years will be given by the formula

$$V = Ae^{rn}$$

By rearranging this formula we can construct an expression for discounting. Suppose we wish to know the net present value of a sum V which will be available in n years. If the instantaneous discount rate is r, then the net present value (A) of V is given by the formula

$$A = Ve^{-rn}$$

The exponential function is intimately related to the *logarithmic* function : if $y = e^x$ is the exponential function then the logarithmic function $y = \log_e x$ is its inverse. A *logarithm* is the *power* to which a base must be raised to obtain a particular number. Thus, $\log_3 9 = 2$ or the base 3 raised to the power of 2 equals the number 9 and $\log_e 1 = 0$ or e raised to the power of 0 equals 1. The natural logarithm, defined as $\log_e$, is commonly written as "ln" and is only defined in the domain of strictly positive numbers. For a logarithm with a base b, for every positive x it is the case that $b^{\log_b x} = x$. Useful properties of logarithms include:

1. $\ln(xy) = \ln x + \ln y$

2. $\ln \frac{x}{y} = \ln x - \ln y$

3. $\ln x^a = a \ln x$

where the variables x and y must be positive.

Another useful result is that the point elasticity of a function $y = f(x)$ of y with respect to x, defined as E_x, where both y and x are positive variables, is the logarithmic derivative, i.e.,

$$E_x = \frac{d(\ln y)}{d(\ln x)} = \frac{dy}{dx} \frac{x}{y}$$

There a number of rules concerning the derivative of a combination of several functions. The first concerns a function of a function and is called the *chain rule*:

$$\frac{df(g(x))}{dx} = \frac{df(g(x))}{dg(x)} \cdot \frac{dg(x)}{dx}.$$

For example

$$\frac{d\ln(2x)}{dx} = \frac{d\ln(2x)}{d(2x)} \cdot \frac{d(2x)}{dx}$$

$$\Rightarrow \frac{d\ln(2x)}{dx} = \frac{1}{2x} \cdot 2 = \frac{1}{x}.$$

Two other useful rules of differentiation are the *product rule*:

$$\frac{d(f(x) \cdot g(x))}{dx} = g(x) \cdot \frac{df(x)}{dx} + f(x) \cdot \frac{dg(x)}{dx}.$$

and the *quotient rule*

$$\frac{d}{dx} \frac{f(x)}{g(x)} = \frac{\frac{df(x)}{dx} g(x) - f(x) \frac{dg(x)}{dx}}{(g(x))^2}.$$

The rules of differentiation can also be applied to functions of several variables. The change in the dependent variable (y) at a point from a change in *one* of the independent variables (x_1), holding all other variables fixed, is called the *partial derivative* and is written as $\frac{\partial y}{\partial x_1}$. Thus, for the function $y = 2x_1^3 + 3x_1 x_2$, the partial derivative is

$$\frac{\partial y}{\partial x_1} = 6x_1^2 + 3x_2.$$

A function $f(x_1, x_2, \cdots, x_n)$ is *continuously differentiable* n times if all the partial derivatives exist up to n times and are continuous.

A useful result in partial differentiation is *Young's Theorem*, which states that provided a function is twice continuously differentiable, the cross-partial derivatives are invariant to the order of differentiation. Thus, if the function $f(x_1, x_2)$ is twice continuously differentiable, then $f_{12} = f_{21}$.

Another important concept is that of a *differential*. For a function $y = f(x)$ we can denote a change in x and y by the differential of x (dx) and the differential of y (dy). Thus, the differential of the function $y = f(x)$ is

$$dy = f'(x)dx$$

The differential dy represents the change in y that would occur if x were changed by an amount equal to dx at the fixed rate $f'(x)$. It, therefore, approximates the change in y from a change in x defined by dx. In the case where the function has two or more independent variables we can apply the notion of a *total differential*. Thus, the total differential of $y = f(x_1, x_2, \ldots, x_n)$ is

$$dy = \frac{\partial f(x)}{\partial x_1}dx_1 + \frac{\partial f(x)}{\partial x_2}dx_2 + \ldots + \frac{\partial f(x)}{\partial x_n}dx_n$$

A useful result that uses derivatives is the *implicit function theorem* which permits us to solve the derivatives of a continuous differentiable function that cannot be written explicitly. The theorem states that given $F(y, x_1, x_2, \ldots x_n) = 0$ and if the implicit function $y = f(x_1, x_2 \ldots x_n)$ exists and is continuously differentiable then its partial derivatives are:

$$\frac{\partial y}{\partial x_i} = -\frac{\frac{\partial F}{\partial x_i}}{\frac{\partial F}{\partial y}}$$

provided that $\frac{\partial F}{\partial y} \neq 0$. For a system of equations (not necessarily linear) one can construct the *Jacobian* matrix which is the matrix of the first order partial derivatives of the system. For example, for the system:

$$\begin{aligned} F_1(x_1, x_2, y_1, y_2) &= 0 \\ F_2(x_1, x_2, y_1, y_2) &= 0 \end{aligned}$$

where x_1 and x_2 are the *exogenous* variables treated as given and y_1 and y_2 are the *endogenous* variables for which we must solve. The Jacobian matrix is

$$\mathbf{J} = \frac{\partial \mathbf{F}}{\partial \mathbf{y}} = \begin{pmatrix} \partial F_1/\partial y_1 & \partial F_1/\partial y_2 \\ \partial F_2/\partial y_1 & \partial F_2/\partial y_2 \end{pmatrix}.$$

More generally, if there are n continuously differentiable functions and n endogenous variables, the Jacobian will be an $n \times n$ matrix where a row i consists of the partial derivatives of function i

with respect to the n endogenous variables. A test for functional dependence among the n equations is to calculate the *Jacobian determinant* often defined as $\det(\mathbf{J})$ or $|\mathbf{J}|$. If the determinant equals zero for all values of y then the equations are functionally dependent and it is not possible to define the endogenous variables solely in terms of the exogenous variables. Conversely, if $|\mathbf{J}| \neq 0$ such that the Jacobian is non-singular (see chapter 2) and if each $\mathbf{F}_i$ is continuously differentiable, then it is possible to define the endogenous variables in terms of the exogenous variables in a neighbourhood of the *equilibrium point* (x_0, y_0) which satisfies the original system of equations. The usefulness of the Jacobian matrix is that it permits us to solve the partial derivatives of a system of nonlinear but continuously differentiable functions.

Another result that is widely applied in economics is the *Taylor series*. This involves approximating the value of a function by the use of a *polynomial* where the polynomial of a variable x has the following form:

$$a_0 + a_1 x + a_2 x^2 + \ldots + a_n x^n$$

Each a_i term is a coefficient while each x term is raised to a non-negative integer. The degree of the polynomial refers to the highest power to which the variable is raised. All polynomial functions are continuous functions. The *Taylor series approximation* of the function $f(x)$ at $x = x_0$ where x_0 is any real number is defined as:

$$f(x) \approx f_n(x) = a_0 + a_1(x - x_0) + a_2(x - x_0)^2 + \ldots + a_n(x - x_0)^n$$

where $a_0 = f(x_0), a_1 = f'(x_0), a_2 = \frac{f''(x_0)}{2!}, \ldots a_n = \frac{f^n(x_0)}{n!}$ and the symbol ! is the *factorial* where $n!$ is defined as $1 \times 2 \times 3 \times \ldots n$.

Another important result is the *intermediate value theorem*. The theorem states that if a continuous function $g(x)$ takes on the value $g(a)$ at a, and $g(b)$ at b where $g(a) < 0 < g(b)$ in a closed interval $[a, b]$ then there exists a point c in the open interval (a, b) such that $g(c) = 0$. More generally, if $g(x)$ is continuous on $[a, b]$ then $g(x)$ must assume every value between $g(a)$ and $g(b)$ at some point x in the interval (a, b).

A concept particularly useful in mathematical economics is that of *concavity*. The formal definition is as follows. A function $f(x_1, \ldots, x_n)$ is concave if for all $\mathbf{x} = x_1, \ldots, x_n$ in the *domain* of the function (the set of all possible values of $\mathbf{x}$) defined by D it is the case that:

$$\lambda \cdot f(\mathbf{x}) + (1 - \lambda) \cdot f(\mathbf{y}) \leq f(\lambda \cdot \mathbf{x} + (1 - \lambda) \cdot \mathbf{y}),$$

where $y \in D$ and $\lambda \in [0,1]$. In other words, the area below the graph of the function must be a convex set, as in the diagram below. If the above inequality holds strictly for all $\lambda \in (0, 1)$ the function is strictly concave.

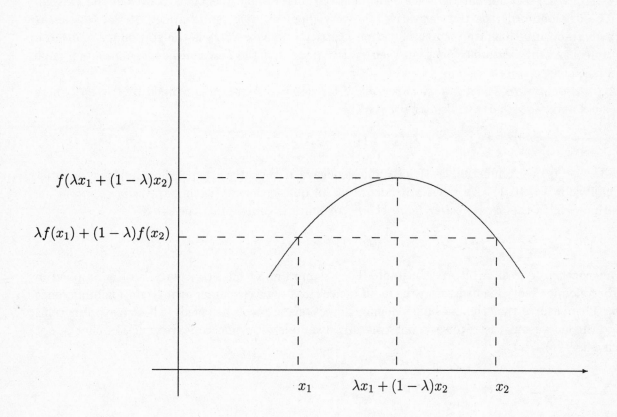

If a function is twice continuously differentiable then it is concave if the Hessian matrix of second-order partial derivatives, defined below, is *negative semidefinite* in the entire domain of the function. It is strictly concave if the Hessian is *negative definite* in the entire domain of the function. If x_0 is the stationary point of a strictly concave differentiable function, then x_0 is unique and the function has its maximum value at x_0 (see chapter 4).

$$\mathbf{H(x)} = \begin{pmatrix} \frac{\partial^2 f(\mathbf{x})}{\partial x_1^2} & \cdots & \frac{\partial^2 f(\mathbf{x})}{\partial x_1 \cdot \partial x_n} \\ \vdots & & \vdots \\ \frac{\partial^2 f(\mathbf{x})}{\partial x_n \cdot \partial x_1} & \cdots & \frac{\partial^2 f(\mathbf{x})}{\partial x_n^2} \end{pmatrix}.$$

The matrix is positive definite, if **all** the leading principal minors are positive and is negative definite if $(-1)^i |\mathbf{D}_i| > 0$ for $i = 1, 2, \cdots n$ where $|\mathbf{D}_i|$ is the ith leading principal minor obtained from a matrix of order i. The matrix is positive semidefinite if **all** the principal minors, not just the leading principal minors, are ≥ 0 and is negative semidefinite if $(-1)^i |\mathbf{B}_i| \geq 0$ for all $i = 1, 2, \cdots n$ where $|\mathbf{B}_i|$ is a principal minor obtained from a matrix of order i (see chapter 2).

A related concept is the *convex* function, where the area *above* the graph is a convex set. If a function $f(x)$ is concave then $-f(x)$ is convex. A function $f(x_1, \ldots, x_n)$ is convex if for all $\mathbf{x} = x_1, \ldots, x_n$ in the *domain* of the function (the set of all possible values of $\mathbf{x}$) defined by D it is the case that:

$$\lambda \cdot f(\mathbf{x}) + (1 - \lambda) \cdot f(\mathbf{y}) \geq f(\lambda \cdot \mathbf{x} + (1 - \lambda) \cdot \mathbf{y}),$$

where $y \in D$ and $\lambda \in [0,1]$. If the above inequality holds strictly for all $\lambda \in (0,1)$ the function is strictly convex. If a function is twice continuously differentiable then it is convex if the Hessian matrix of second-order partial derivatives is everywhere *positive semidefinite*. If the Hessian is everywhere positive definite then the function is strictly convex. If x_0 is the stationary point of a strictly convex differentiable function, then x_0 is unique and the function has its minimum value at x_0 (see chapter 4).

A function $f(x)$ is said to be *quasiconcave* if for *any* two values of x defined by (x_1, x_2) where $f(x_1) = c$ and $f(x_2) \geq c$ it is also true that

$$f(\lambda x_1 + (1 - \lambda)x_2) \geq c$$

for all $\lambda \in [0,1]$. A function is strictly quasiconcave if the above expression holds as a strict inequality for $\lambda \in (0,1)$. An alternative definition of quasiconcavity is that the *upper contour set* sometimes called the *weakly better than* set in economics is *convex*, i.e., the set S

$$S = \{x : f(x) \geq c\},$$

is convex where c is an arbitrary constant. The assumption of quasiconcavity is often applied in microeconomic theory to ensure that the indifference curves of a consumer form a convex indifference map. This ensures that the consumer demand functions are well behaved. All concave functions are quasiconcave but the reverse is not true. An example of a quasiconcave function that is not concave is $f(x) = x^3$.

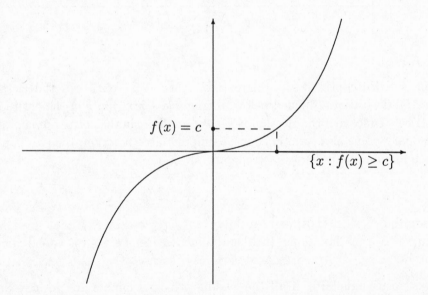

From the diagram above we can see that the set of x values for which $f(x) \geq c$ (where c is an arbitrary value) is all x on the darkly shaded line on the x axis. Whichever two points we pick on this shaded line, if we draw a line between the two points, all the points on the line are in the set S such that $f(x) > c$. Thus, the set is convex and because this is true for **any** value of c that we pick, the function is quasiconcave.

A function $f(x)$ is said to be *quasiconvex* if for any two values of x defined by (x_1, x_2) where $f(x_1) = c$ and $f(x_2) \leq c$ it is also true that

$$f(\lambda x_1 + (1 - \lambda)x_2) \leq c$$

for all $\lambda \in [0,1]$. A function is strictly quasiconvex if the above expression holds as a strict inequality and $\lambda \in (0, 1)$. An alternative definition of quasiconvexity is that the *lower contour* set, sometimes known as the *weakly worse than* set in economics, is *convex*, i.e., the set S

$$S = \{x : f(x) \leq c\},$$

is convex where c is an arbitrary constant. All convex functions are quasiconvex but the reverse is not true.

If a function is twice continously differentiable and is defined in an open, convex set (such as the non-negative orthant where $x_1, x_2, \cdots, x_n > 0$) a determinant test can be used to verify whether it is quasiconcave or quasiconvex. It is necessary to form a matrix which consists of the Hessian matrix with a border of first order partial derivatives defined by $\mathbf{B}$.

$$\mathbf{B}(\mathbf{x}) = \begin{pmatrix} 0 & \frac{\partial f(\mathbf{x})}{\partial x_1} & \cdots & \frac{\partial f(\mathbf{x})}{\partial x_n} \\ \frac{\partial f(\mathbf{x})}{\partial x_1} & \frac{\partial^2 f(\mathbf{x})}{\partial x_1^2} & \cdots & \frac{\partial^2 f(\mathbf{x})}{\partial x_1 \cdot \partial x_n} \\ \vdots & & & \vdots \\ \frac{\partial f(\mathbf{x})}{\partial x_n} & \frac{\partial^2 f(\mathbf{x})}{\partial x_n \cdot \partial x_1} & \cdots & \frac{\partial^2 f(\mathbf{x})}{\partial x_n^2} \end{pmatrix}.$$

A necessary condition for quasiconcavity is that for all values of x in the domain of the function $(-1)^i |\mathbf{D}_i| \geq 0$ for $i = 1, 2, \cdots n$ where $|\mathbf{D}_i|$ is the $i+1$ leading principal minor while a sufficient condition for quasiconcavity is $(-1)^i |\mathbf{D}_i| > 0$ for $i = 1, 2, \ldots n$. A necessary condition for quasiconvexity is that for all values of x in the domain of the function $|\mathbf{D}_i| \leq 0$ for $i = 1, 2, \ldots n$ while a sufficient condition for quasiconvexity is $|\mathbf{D}_i| < 0$ for $i = 1, 2, \ldots n$. All single variable functions that are strictly (monotonically) increasing or decreasing functions are both quasiconcave and quasiconvex.

Homogeneity is another concept that is particularly useful in economics. A function $f(x_1, \ldots, x_n)$ is homogeneous of degree k if $f(\lambda x_1, \ldots, \lambda x_n) = \lambda^k \cdot f(x_1, \ldots, x_n)$, for all $\lambda > 0$. Thus, if a production function is homogeneous of degree one (linearly homogeneous), then a doubling of all inputs will result in a doubling of the output. An important characteristic of homogeneous functions is that the level curves of such functions, such as the indifference curves for a utility function or the isoquants of a production function, are radial blowups of each other. This means that the expansion paths are straight lines.

This property is also shared by *homothetic* functions. More precisely, the ratio of the partial derivatives is invariant to the scaling of the arguments of the function. Consider the function $f(x, y)$. Suppose that at the point (x_1, y_1) the ratio of the partial derivatives is

$$\frac{\frac{\partial f}{\partial x}}{\frac{\partial f}{\partial y}} = c.$$

If the function is homethetic, the ratio of partial derivatives will equal c for any point (x, y) along the same ray from the origin as (x_1, y_1), i.e., any point (x, y) such that

$$(\lambda x, \lambda y) = (x_1, y_1)$$

for some $\lambda > 0$.

Homothetic functions are, in general, not homogeneous but are strictly increasing (monotonic) transformations of homogeneous functions. Thus, if $y = f(x_1, x_2)$ is homogeneous of degree k provided that $\frac{dV}{dy} > 0$ then the function $V(f(x_1, x_2))$ is homothetic. In other words, a function is homothetic if it can be expressed as a strictly increasing function $V()$ of a homogeneous function. It follows that homogeneous functions are a subset of homothetic functions.

A result related to the concept of homogeneity is *Euler's Theorem* which states that for a function $f(x_1, \ldots, x_n)$ homogeneous of degree k,

$$\frac{\partial f}{\partial x_1} x_1 + \frac{\partial f}{\partial x_2} x_2 \ldots \frac{\partial f}{\partial x_n} x_n \equiv k f(x_1, x_2, \ldots x_n)$$

Thus, for a function $f(x_1, \ldots, x_n)$ homogenous of degree $k = 0$,

$$\frac{\partial f}{\partial x_1} x_1 + \frac{\partial f}{\partial x_2} x_2 \ldots \frac{\partial f}{\partial x_n} x_n \equiv 0$$

Further Reading

For readers seeking a refresher in algebra and how to solve equations we suggest Haeussler and Paul [21] (chapters 1 and 2). We strongly recommend Sydsæter and Hammond [32] (chapters 4 and 5), Dowling [17] (chapters 3-5) and Toumanoff and Nourzad [34] (chapters 3-5) if you need additional practice in differentiation and applications to economics. A good introduction to the ideas used in this chapter is Glaister [19] (chapters 11 to 14). Other accessible texts include Birchenall and Grout [6] (chapters 1 to 4), Archibald and Lipsey [1]. Bressler [10] (chapter 6) also provides a good review of limits and their application in economics. Ostrosky and Koch [25] (chapter 3) present a nice review of the concepts of limits, differentiability and continuity as do Sydsæter and Hammond [32] (chapters 6 and 7). Haeussler and Paul [21] provide very readable introductions to limits and continuity (chapter 11) and differentiation (chapters 12, 13 and 15).

A slightly more advanced treatment on functions and sets is provided by Sydsæter [31] (chapters 3 and 5), which provides a detailed examination of the implicit function theorem. A very helpful book on sets and functions is Binmore [5]: although written for second year mathematics students this is a useful reference for those with the requisite skills. Finally, for advanced students a standard and authoritative reference on real analysis is Rudin [27].

Chapter 3 - Questions

Sets

Question 1

Determine whether the following sets are open, closed, bounded, and/or convex. (Hint: Try to sketch the sets).

(i) $\mathcal{R}^2_{++} = \{\mathbf{x} = (x_1, x_2) \in \mathcal{R}^2 : x_1 > 0, x_2 > 0\}$.

(ii) $S = \{\mathbf{x} \in \mathcal{R}^2 \mid 0 < d(\mathbf{x}, \mathbf{0}) \leq r\}$, where $d(\mathbf{x}, \mathbf{y})$ is the distance between two points in $\mathcal{R}^2$, which we define by

$$d(\mathbf{x}, \mathbf{y}) = \sqrt{(x_1 - y_1)^2 + (x_2 - y_2)^2}.$$

(iii) $S = \mathcal{R}^2$

(iv) $S = \{\mathbf{x} \in \mathcal{R}^2 : \mathbf{x} \in \mathcal{R}^2_+ \text{ and } x_2 \leq \frac{1}{x_1}\}$

Natural Exponential Function

Question 2

Suppose an individual invests \$100 for one year in a bank account paying 10% per year nominal interest.

(i) If the interest was payable every 6 months how much would the individual have in principal and interest at the end of 12 months?

(ii) If the number of payments in interest made in the 12 month period equals n, how much would the individual have in principal and interest at the end of 12 months if $n \to \infty$?

Question 3

What is the net present value today of \$100 in ten years time, given an instantaneous discount rate of 5%?

Logarithms

Question 4

(i) Suppose the demand function of a good x is defined as

$$D(p) = Ap^{-b}$$

where A, p and $b > 0$.

Calculate the price elasticity of demand defined as $\frac{p}{D(p)} \frac{dD(p)}{dp}$.

(ii) Prove that for the function $y = f(x)$ where both y and x are positive variables that the point elasticity of y with respect to a change in x (ED_x) is the logarithmic derivative, i.e.,

$$ED_x = \frac{d(\ln y)}{d(\ln x)}$$

Limits

Question 5

Use L'Hôpital's rule to evaluate the following limits

(i)

$$\lim h(x) \text{ as } x \to 0, \text{ where } h(x) = \frac{3x^2}{e^x - 1}.$$

(ii)

$$\lim h(x) \text{ as } x \to -1, \text{ where } h(x) = \frac{x + 1}{x^2 - 1}.$$

The Intermediate Value Theorem

Question 6

(i) Determine that a solution exists for the following fourth order equation in the interval $[-2, 3]$.

$$x^4 + x^2 + 10.5x = 0$$

(ii) Suppose that a firm has the following profit function:

$$\Pi(L) = f(L) - WL, W > 0.$$

where $f(L)$ is continuously differentiable on $\mathcal{R}_{++}$, $f(L) > 0$ for $L > 0$, $\lim_{L \to 0} f'(L) = \infty$, $\lim_{L \to \infty} f'(L) = 0$ and $f''(L) < 0$.

Show that there is a profit maximizing value of L, $L^* > 0$, that maximizes $\Pi(L)$. Is L^* unique? If so explain why, if not explain what condition would be required.

Continuity

Question 7

Where are the following functions continuous, and where are they discontinuous?

(i) $f(x) = \begin{cases} 1, x = 0, \\ 0, \text{ otherwise} \end{cases}$

(ii) $g(x) = \frac{1}{x^2+1}$.

(iii) $h(x) = f(g(x)), x > 0$.

Differentiability

Question 8

Are the following functions differentiable at zero?

(i) $f(x) = \text{abs}(x)$, where $\text{abs}(\pm x) = +x$. (e.g. $|-5| = 5$).

(ii) $f(x) = (\text{abs}(x))^2$.

(iii) $f(x) = \begin{cases} 1, x \leq 0 \\ 0, x > 0 \end{cases}$

Question 9

Differentiate the following functions with respect to x.

(i) $f(x) = e^{x^2}$

(ii) $g(x) = \frac{\sqrt{x}}{3x^2}$

(iii) $h(x, y) = x^2 + y^2 + (x + y)^2$

Increasing Functions

Question 10

Determine whether the following functions are increasing, and if so whether or not they are strictly increasing.

(i) $f(x) = \ln(x + 1)$, where $x > -1$

(ii) $f(x) = -\alpha e^{-\alpha x}$, where $\alpha \neq 0$

(iii) $f(x) = x^3$, where $x \in \mathcal{R}$

Concavity and Convexity

Question 11

Determine whether the following functions are concave, convex, quasiconcave or quasiconvex.

(i)

$$y = \ln(x+1)$$

where $x > 0$

(ii)

$$y = x_1^2 + 2x_2^2$$

(iii)

$$y = 6$$

(iv)

$$y = x_1 x_2 + x_1 + x_2$$

where $x_1, x_2 > 0$

(v)

$$y = \frac{x_1^2}{2} + \frac{x_2^2}{2}$$

(vi)

$$y = -e^{-x} + 12x$$

(vii)

$$f(x_1, x_2, x_3) = x_1^3 + x_2^3 + x_3^3 + 3x_1 + 3x_2 + 3x_3$$

where $x_1, x_2, x_3 > 0$

Homogeneity and Homotheticity

Question 12

Where possible calculate the degree of homogeneity of the following functions:

(i) $f(x, y) = A \cdot x^\alpha \cdot \sqrt{y}$

(ii) $f(x, y) = \frac{xy}{x+y}$

(iii) $f(x, y) = \ln(x + y)$

(iv) $f(x, y, z) = \ln(x^2 + y^2 + z^2) + 2 \cdot \ln\left(\frac{2}{x} + \frac{3}{y} + \frac{4}{z}\right)$

Question 13

Which of the following functions are homothetic?

(i) $f(x) = x^2 y^4$

(ii) $f(x, y) = e^{x^2 + y^2}$

(iii) $(x^2 - 1)^2$

Euler's Theorem

Question 14

Using Euler's Theorem show that for the following production function that if the average product of labor $(\frac{Q}{L})$ is rising, then the marginal product of capital $(\frac{\partial Q}{\partial K})$ is falling.

$$Q = K^\alpha L^{1-\alpha}$$

Implicit Function Theorem

Question 15

Which of the following equations define y as a function of the x's (i.e., when is it the case that for each possible value of x, there is only one value of y that solves the equation). For those cases where y is a function of x, find the derivative of y (where it exists).

(i) $(y - x)^2 = 1$

(ii) $(y - x)^2 = -1$

(iii) $\ln(x + y) - x + y = 1, x, y > 0$

(iv) $e^{x_1 + y} - x_2 + y = 1$

Question 16

For the following system of equations, calculate the partial derivatives of the dependent variables (y_1, y_2) with respect to the independent variables (x_1, x_2). Assume that the equations are consistent for at least some values of the variables.

$$
\begin{aligned}
F_1(\mathbf{y}, \mathbf{x}) &= e^{y_1 + x_1} + y_2 + x_2 = 0 \\
F_2(\mathbf{y}, \mathbf{x}) &= y_2 + 2x_2 - x_1 = 0
\end{aligned}
$$

Taylor Series

Question 17

(i) Let $f(x) = \ln(x)$

Calculate the first order Taylor expansion to f around $x = 1$ and then find the first order Taylor series approximation to $f(1 + r)$ where $r > 0$.

(ii) Let $f(x) = x^2(x - 2)^2$

Calculate the first order Taylor approximation to f around $x_0 = 0$, and so provide estimates of $f(1)$ and $f(2)$. How big is the error from the approximation in each case?

Chapter 3 - Solutions

Sets

Question 1

(i)

The easiest way of determining whether a set is open or not is to identify the boundary of the set and then to use the fact that an open set is one which does not contain its boundary. In the present case the boundary is

$$\{\mathbf{x} \in \mathcal{R}^2 : x_1 = 0, x_2 \geq 0\} \cup \{\mathbf{x} \in \mathcal{R}^2 : x_2 = 0, x_1 \geq 0\}.$$

The symbol $\cup$ means the *union* of two sets, which implies that the boundary set is the collection of all of the elements from both the sets. None of the boundary points are in $\mathcal{R}^2_{++}$ so the set is open. The set is not closed for the same reason, since closed sets are those that contain *all* their boundary points.

A bounded set is one for which the distance between any pair of points is not greater than some finite number, r. Consider the points $(1, r + 2)$ and $(1,1)$. Both these points are in $\mathcal{R}^2_{++}$, however, the distance between them is:

$$\sqrt{(1-1)^2 + (r+2-1)^2} = \sqrt{0 + (r+1)^2} = r + 1 > r$$

Thus, the set is not bounded as the distance $d > r$.

A set is defined to be *convex* if for any two points $\mathbf{x}$ and $\mathbf{y}$ in the set,

$$\mathbf{z}(\lambda) = \lambda \cdot \mathbf{x} + (1 - \lambda) \cdot \mathbf{y} \in S$$

for all $\lambda \in (0,1)$. Expanding $\mathbf{z}$ we obtain

$$\mathbf{z}(\lambda) = (\lambda \cdot x_1 + (1 - \lambda) \cdot y_1, \lambda \cdot x_2 + (1 - \lambda) \cdot y_2).$$

Since x_1, x_2, y_1 and y_2 are all > 0, then so must be $\lambda \cdot x_1 + (1 - \lambda) \cdot y_1$ and $\lambda \cdot x_2 + (1 - \lambda) \cdot y_2$, since λ and $(1 - \lambda)$ are greater than zero. Thus $\mathbf{z}(\lambda) \in \mathcal{R}^2_{++}$ and so it is convex.

(ii)

This expression defines a circle centred about (0,0) with a "hole" at the middle the size of a single point. This set is drawn in the diagram below:

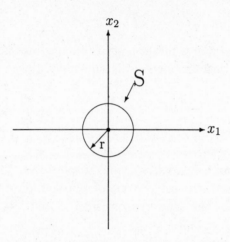

The boundary of this set is denoted by Z, where:

$$Z = (0,0) \cup \{\mathbf{x} \in \mathcal{R}^2 \mid d(\mathbf{x}, 0) = r\}$$

Since (0,0) is an element of Z but not of S the set cannot be closed. Similarly, because $\{\mathbf{x} \in \mathcal{R}^2 \mid d(\mathbf{x}, 0) = r\}$ is in S as well as Z the set cannot be open.

S is bounded, since $d(\mathbf{x}, \mathbf{y}) \leq r$ for all points $\mathbf{x}$ and $\mathbf{y}$ in S. However, S is not convex. Consider the points (1,1) and (-1,-1). Then for $\lambda = \frac{1}{2}$, one has

$$\mathbf{z}(\frac{1}{2}) = (\frac{1}{2} \cdot (1 + -1), \frac{1}{2} \cdot (1 + -1)) = (0,0) \notin \text{S}$$

In other words, since the point (0,0) is halfway along a line drawn from (1,1) and (-1,-1), the set is not convex.

(iii) $S = \mathcal{R}^2$

In this example the set consists of the whole space $\mathcal{R}^2$. This set has no boundary points and is thus unbounded: however this fact makes difficult the task of deciding whether the set is open or closed. This is because we cannot use the result that an open set is one that does not contain its boundary points. Instead we will have to rely upon a more basic definition: a set S is open if for any point $x \in S$, an epsilon ball centred at that point is completely contained within the set S, for a ball small enough. In mathematical terms, it must be that: $\forall \mathbf{x} \in S, B_\epsilon(\mathbf{x}) \in S$, for some ϵ. This must be true for *all* epsilon: no matter how large we draw the ball, it is still contained in $\mathcal{R}^2$.

The set $\mathcal{R}^2$ is closed. To see this one must take an indirect approach. Suppose $\mathcal{R}^2$ were not closed: then there must be some point $\mathbf{x}$ which is a boundary point of the set but which is not in the set. But since S is the whole space, the point $\mathbf{x}$ must be in S. (This is an example of *proof by contradiction*, since we start out by assuming the opposite of what we want to show and then keep going until we arrive at a contradiction.)

The set S is convex because any line drawn between two points will be in $\mathcal{R}^2$.

(iv) $S = \{\mathbf{x} \in \mathcal{R}^2 : \mathbf{x} \in \mathcal{R}^2_+ \text{ and } x_2 \leq \frac{1}{x_1}\}$

This set consists of that part of the non-negative orthant that is underneath the function $x_2 = 1/x_1$. The set is illustrated below:

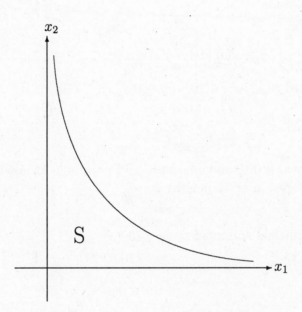

The boundary set for S is made up of the positive parts of the x_1 and x_2 axes, along with the line $x_2 = 1/x_1$. The set is not open, since it contains some of its boundary points. Because S contains all its boundary points it is closed. However the set is not bounded because it contains the whole of the positive part of the x_1 axis, which is infinitely long. Bounded sets cannot contain unbounded sets.

There are several ways to show that S is not convex. One way is to note that $x_2 = 1/x_1$ is not a concave function, thus by definition the area underneath the curve cannot be convex. Alternatively we can proceed from the definition of convexity given above: if we choose the points $\mathbf{x} = (3, 0)$ and $\mathbf{y} = (0, 3)$, then the point $\mathbf{z} = (1.5, 1.5)$ is not in S since

$$z_2 = 3/2 > 1/z_1 = 2/3.$$

Natural Exponential Function

Question 2

(i)

If the individual were paid interest every six months the amount she would have at the end of 12 months in principal and interest would be

$$100 + 100(0.05) + (100(1.05))0.05 \; = \; 110.25$$

where the first term is the principal, the second term is the interest received after 6 months and the third term is the interest received after 12 months. By increasing the frequency of the interest payment the individual benefits by receiving interest on the interest previously received.

(ii)

If the number of interest payments in the 12 month period approaches infinity we can use the natural exponential function to determine the principal and interest due to the individual after 12 months.

The number e is defined as follows

$$e \equiv \lim_{n \to \infty} (1 + \frac{1}{n})^n$$

It is also true that e^x is defined as

$$e \equiv \lim_{n \to \infty} (1 + \frac{x}{n})^n$$

Thus, if the interest rate per year is 0.10 and interest is paid n times in 12 months where $n \to \infty$ the value of the principal and interest after 12 months is

$$\begin{aligned} \text{principal and interest} \quad &= \quad 100(e^{0.1}) \\ &= \quad 110.52 \end{aligned}$$

Question 3

The answer can be found from the formula

$$A = V(e^{-rn})$$

In this case, the time period n is 10, the discount rate r is 0.05, and the sum of money V is \$100. Thus, the net present value is

$$A = 100e^{-0.05(10)} = 60.65.$$

Logarithms

Question 4

(i)

The elasticity can be calculated using derivatives such that

$$\begin{aligned} ED_p \quad &= \quad \frac{p}{D(p)} \frac{dD(p)}{dp} \\ &= \quad \frac{p}{Ap^{-b}} -bAp^{-b-1} \\ &= \quad -b \end{aligned}$$

The result could also be found using logarithms. Taking the logarithm on both sides of $x = D(p) = Ap^{-b}$ we obtain

$$\begin{aligned} \ln x &= \ln A + \ln(p^{-b}) \\ &= \ln A - b \ln p \end{aligned}$$

Taking the differential of both sides of the equation noting that $\frac{d\ln(x)}{dx} = \frac{1}{x}$ and that A is a constant we obtain

$$\frac{dx}{x} = -b\frac{dp}{p}$$

From the differential we can state that the percentage change in x is $-b$ multiplied by the percentage change in p. In other words, $-b$ is the price elasticty of demand of the good x or the percentage change in x due to a percentage change in p, i.e.,

$$ED_p = \frac{dx}{dp}\frac{p}{x} = -b$$

(ii)

In general, the point elasticity of a function $y = f(x)$ of y with respect to x where both y and x are positive variables is

$$ED_x = \frac{d(\ln y)}{d(\ln x)}$$

This can be shown as follows noting that $\ln y$ is a differentiable function of y which itself is a function of x. We may also note that $e^{\ln x} = x$ thus x can be defined as a function of $\ln x$. The chain rule of differentiation states that if $f()$ is a function of $g()$ which itself is a function of x then

$$\frac{df(g(x))}{dx} = \frac{df(g(x))}{dg(x)} \cdot \frac{dg(x)}{dx}.$$

Applying it to the equation for the point elasticity and noting that $\frac{de^x}{dx} = e^x$ we obtain

$$\begin{aligned} \frac{d\ln y}{d\ln x} &= \frac{d\ln y}{dy} \cdot \frac{dy}{dx} \cdot \frac{dx}{d\ln x} \\ &= \frac{1}{y} \cdot \frac{dy}{dx} \cdot \frac{de^{\ln x}}{d\ln x} \\ &= \frac{1}{y} \cdot \frac{dy}{dx} \cdot e^{\ln x} \\ &= \frac{1}{y} \cdot \frac{dy}{dx} \cdot x \\ &= \frac{dy}{dx} \cdot \frac{x}{y} \end{aligned}$$

Limits

Question 5

(i)

L'Hôpital's rule is

$$\lim_{x \to a} \frac{f(x)}{g(x)} = \lim_{x \to a} \frac{f'(x)}{g'(x)}$$

where $g'(x) \neq 0$ for $x \neq a$. The rule is particularly useful when the numerator of the left hand side tends to infinity or the denominator tends to zero, since it may be that the ratio of the *derivatives* will produce a finite quantity.

L'Hôpital's rule is applicable in the present situation because if we define $f(x) = 3x^2$ and $g(x) = e^x - 1$, then $f(0) = 0$, and $g(0) = 1 - 1 = 0$. Further, $g'(x) = e^x > 0$ for all x. The next step is to form the appropriate ratio of derivatives:

$$f'(x) = 6x, g'(x) = e^x$$
$$\Rightarrow \lim_{x \to 0} \frac{f'(x)}{g'(x)} = \frac{0}{1} = 0.$$

(ii)

L'Hôpital's rule is applicable in the present situation because if we define $f(x) = x$ and $g(x) = x^2 - 1$, then $f(-1) = 0$, and $g(-1) = 0$. Further, $g'(-1) \neq 0$ for all x. Forming the appropriate ratio of derivatives we obtain

$$f'(x) = 1, g'(x) = 2x$$
$$\Rightarrow \lim_{x \to -1} \frac{f'(x)}{g'(x)} = \frac{1}{-2}.$$

The Intermediate Value Theorem

Question 6

(i)

We can define the polynomial function as follows

$$f(x) = x^4 + x^2 + 10.5x$$

Because $f(x)$ is a polynomial it is a continuous function and thus the intermediate value theorem is applicable.

At the value of $x = -2$, $f(-2) = -1$ and at the value $x = 3$, $f(3) = 121.5$. Because $f(x)$ takes on every value between -1 and 121.5, including zero, in the interval $(-2, 3)$ it follows that there must be some x in this interval such that $f(x) = 0$. Thus, from the intermediate value theorem the fourth order equation must have a solution or a root in the interval $x \in (-2, 3)$.

(ii)

The first order condition to the firm's problem is

$$\Pi'(L) = f'(L) - W = 0.$$

The intermediate value theorem states that if a continuous function $g(x)$ takes on the value $g(a)$ at a, and $g(b)$ at b, then for all numbers c between $g(a)$ and $g(b)$ there is some point $x \in (a, b)$ such that $g(x) = c$.

In our case we know that $\Pi'(L)$ is continuous since $f(L)$ is continuously differentiable, thus the theorem applies. We have $\Pi'(0) = \infty$ because $\lim_{L \to 0} f'(L) = \infty$ and $\Pi'(\infty) = -W$ because $\lim_{L \to \infty} f'(L) = 0$. Because 0 is between $-W$ and ∞, the intermediate value theorem states that $\Pi'(L^*) = 0$ for some $L^* \in (0, \infty)$. It follows that if $\Pi'(L^*) = 0$ then L^* solves the firm's first order conditions. Because $\Pi''(L) = f''(L) < 0$ the function is strictly concave and the solution is a unique maximum.

Continuity

Question 7

(i)

By sketching the function we can show that the function is discontinuous at 0 because we must take our pen off the paper to mark the value of the function at $x = 0$. There are several equivalent ways of showing the discontinuity at $x = 0$. Perhaps the most intuitive is the following: a function $f(x)$ is continuous at a point z in $\mathcal{R}$ if

$$\lim_{x \to z} f(x) = f(z)$$

Implicit in this definition is the requirement that the limit $f(z)$ actually exists.

How do we apply this definition to the present case? Consider the point $z = 0$. Then $f(z) = f(0) = 1$. We now need to calculate the following limit:

$$\lim_{x \to 0} f(x) = 0$$

No matter how close x is to zero, as long x does not equal zero then $f(x) = 0$. Thus $\lim f(x) \neq f(0)$ and so $f()$ is not continuous at zero.

What about points z other than zero? In this case we have seen that $\lim f(z) = 0$, and we know that $f(z) = 0$ for $z \neq 0$. Thus $\lim f(x) = f(z)$ when x does not equal zero, and so the function is continuous at all points $x \neq 0$.

(ii)

To solve this part of the question we use again the definition above. The first step is to compute the limit of the function around a point z:

$$\lim_{x \to z} \frac{1}{x^2 + 1} = \lim_{x \to z} (x^2 + 1)^{-1} = (z^2 + 1)^{-1}.$$

Here we have used the fact that $\lim(x/y) = \lim(x)/\lim(y)$. Note that the fraction on the right exists for all z, since the denominator is always positive. Because the $f(z)$ also is equal to $\frac{1}{z^2+1}$, the function $g(x)$ is continuous for all x.

(iii)

This function is continuous for all $x > 0$. This can be proved as follows. From (i) and (ii) we know that $f()$ and $g()$ are continuous for $x > 0$. Thus, we can use the rule that a composite function $h = f(g(x))$ is continuous if its component functions $f()$ and $g()$ are continuous.

Differentiability

Question 8

(i)

This function is not differentiable at $x = 0$. To show this we must return to the definition: a function f is differentiable at x_0 if

$$\lim_{\Delta x \to 0} \frac{f(x_0 + \Delta x) - f(x_0)}{\Delta x}$$

exists.

In our case we are interested in the point $x_0 = 0$. Beginning with the left hand limit of the difference quotient

$$\lim_{\Delta x \to 0^-} \frac{f(0 + \Delta x) - f(0)}{\Delta x} = \lim_{\Delta x \to 0^-} \frac{\text{abs}(\Delta x) - 0}{\Delta x}$$

Defining $x = x_0 + \Delta x$ such that $\Delta x = x - x_0$ we obtain

$$\lim_{x \to 0^-} \frac{\text{abs}(x - x_0) - 0}{x - x_0} = -1$$

In this case, $x < 0$, $\text{abs}(x) = -x$ and $x_0 = 0$ such that the left hand limit of the difference quotient is -1. The right hand side limit of the difference quotient is found in a similar manner noting that in this case $x > 0$ such that

$$\lim_{x \to 0^+} \frac{\text{abs}(x - x_0)}{x - x_0} = 1$$

The left and right hand limits of the difference quotients are not equal at 0, thus the limit of the difference quotient does not exist. It follows, therefore, that the function is not differentiable at $x_0 = 0$. This is because the slope of the function for $x > 0$ is positive (+1), and the slope for $x < 0$ is negative (-1).

(ii)

Once again it is easiest to consider left and right hand limits of the difference quotients separately. For the left hand limit:

$$\lim_{\Delta x \to 0^-} \frac{f(0 + \Delta x) - f(0)}{\Delta x} = \lim_{\Delta x \to 0^-} \frac{(\text{abs}(\Delta x))^2 - 0}{\Delta x}$$

Defining $x = x_0 + \Delta x$ such that $\Delta x = x - x_0$ we obtain

$$\lim_{x \to 0^-} \frac{(\text{abs}(x - x_0))^2 - (\text{abs}(0))^2}{x - x_0}$$

Given that $x_0 = 0$ we can rewrite the limit as follows

$$\lim_{x \to 0^-} \frac{x^2 - 0}{x} \Rightarrow \lim_{x \to 0^-} x = 0$$

Using the same procedure for the right hand limit we obtain

$$\lim_{\Delta x \to 0^+} \frac{f(0 + \Delta x) - f(0)}{\Delta x} = \lim_{x \to 0^+} \frac{(x)^2 - 0}{x} = \lim_{x \to 0^+} x = 0$$

Because the values of the left and right hand limits of the difference quotients are the same, the function is differentiable at zero.

(iii)

This function is not differentiable at 0 because it is not continuous at 0: the latter is a necessary though not sufficient condition for differentiability, i.e., a differentiable function must be continuous but a continuous function need not be differentiable. A function $f(x)$ is continuous at $x = x_0$ when x_0 is in the domain of the function, a limit exists at x_0 and $\lim_{x \to x_0} f(x) = f(x_0)$. In this case, $\lim_{x \to 0} f(x) \neq f(0) = 1$.

Question 9

(i) $f(x) = e^{x^2}$

The function f essentially consists of two functions: e^y and x^2. The two simple functions are easy to differentiate, and so the approach to follow is to rewrite $f(x)$ as a combination of the two functions:

$$f(x) = e(y), \quad y(x) = x^2$$

and then to use the chain rule of differentiation. The chain rule implies in the present case that

$$\frac{df(x)}{dx} = \frac{de(y)}{dy} \frac{dx^2}{dx}$$
$$= e^y \cdot 2x = e^{x^2} \cdot 2x$$

(ii) $f(x) = \frac{\sqrt{x}}{3x^2}$

This time we can use the quotient rule of differentiation. Define $g(x) = \sqrt{x}$ and $h(x) = 3x^2$. Then $f(x) = g(x)/h(x)$. The derivatives of $g(x)$ and $h(x)$ are

$$
\begin{aligned}
\frac{dg(x)}{dx} &= \frac{d}{dx}\sqrt{x} = \frac{1}{2}x^{-\frac{1}{2}} \\
\frac{dh(x)}{dx} &= \frac{d}{dx}3x^2 = 6x
\end{aligned}
$$

Inserting these expressions into the chain rule formula gives us

$$
\begin{aligned}
\frac{df(x)}{dx} &= \frac{(\frac{1}{2}x^{\frac{-1}{2}})(3x^2) - (\sqrt{x})(6x)}{9x^4} \\
&= \frac{\frac{3}{2}x^{\frac{3}{2}} - 6x^{\frac{3}{2}}}{9x^4} \\
&= \frac{\frac{-9}{2}x^{\frac{3}{2}}}{9x^4} \\
&= \frac{-x^{\frac{-5}{2}}}{2}
\end{aligned}
$$

(iii) $f(x,y) = x^2 + y^2 + (x+y)^2$

Recall that we want the derivative with respect to x, not y. This means that we want the partial derivative $\partial f(x)/\partial x$. We therefore treat y as a constant. The solution is

$$
\frac{\partial f(x)}{\partial x} = 2x + 2(x+y)
$$

Increasing Functions

Question 10

(i)

Differentiating this function we have:

$$
\frac{df(x)}{dx} = \frac{1}{x+1} > 0.
$$

Thus the function is strictly increasing because its derivative is greater than zero over the whole range.

(ii)

Differentiating the function $f(x) = -\alpha e^{-\alpha x}$, we obtain

$$
\frac{df(x)}{dx} = -\alpha(-\alpha)e^{-\alpha x} = \alpha^2 e^{-\alpha x} > 0.
$$

Again, the function is strictly increasing because its derivative is greater than zero over the whole range.

(iii)

The first derivative of this function is

$$\frac{df(x)}{dx} = 3x^2 \geq 0 \text{ for all } x \in \mathcal{R},$$

$$= 0 \text{ for } x = 0.$$

Thus, one can assert that the function is increasing. Is this true in the strict sense? Despite the zero value for the derivative at $x = 0$, the answer is that the function **is** strictly increasing. This is because the derivative test for a strictly increasing function is a *sufficient* but not a *necessary* condition. A necessary and sufficient condition for a function to be strictly increasing is that for all points a and b in $\mathcal{R}$ where $a < b$ that

$$f(a) < f(b)$$

Set $a = 0$. Then one has to show that

$$b^3 > 0 \text{ if } b > 0.$$

This is always true, thus, f is strictly increasing.

In the diagram below we illustrate the function f. It should be clear that the presence of a stationary point at 0 (a point where $f'(x) = 0$) does not necessarily mean that the function is not strictly increasing.

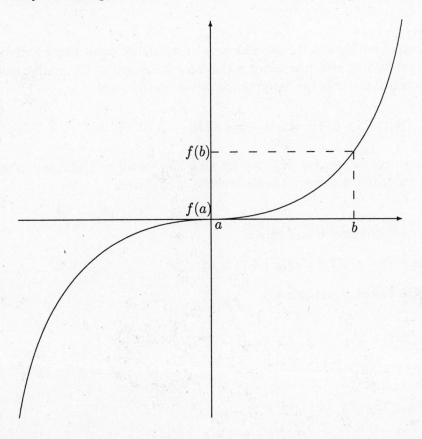

Concavity and Convexity

Question 11

(i)

$$\frac{dy}{dx} = \frac{1}{x+1}$$

$$\frac{d^2y}{dx^2} = \frac{-1}{(x+1)^2} < 0$$

This satisfies the definition of strict concavity as the second derivative of the function is strictly negative for all positive values of x. Because the function is strictly concave it is also strictly quasiconcave. It also satisfies the sufficient condition for a strictly increasing function so it is also quasiconvex.

(ii)

We must form the Hessian matrix of second-order partial derivatives and determine its definitness. The general Hessian matrix is

$$\mathbf{H} = \begin{bmatrix} \frac{\partial^2 y}{\partial x_1^2} & \frac{\partial^2 y}{\partial x_1 \cdot \partial x_2} \\ \frac{\partial^2 y}{\partial x_2 \cdot \partial x_1} & \frac{\partial^2 y}{\partial x_2^2} \end{bmatrix}$$

for the problem at hand we have

$$\mathbf{H} = \begin{bmatrix} 2 & 0 \\ 0 & 4 \end{bmatrix}$$

the successive leading principal minors are the determinants of square matrices formed along the main diagonal where the first leading principal minor is the top left element of the matrix and the second and last leading principal minor is the determinant of the matrix itself.

$$|\mathbf{H_1}| = 2 > 0, |\mathbf{H}| = 2 \cdot 4 - 0 = 8 > 0$$

Thus, the Hessian is positive definite because $|\mathbf{H_n}| > 0$ for all n and for all x. The function is, therefore, strictly convex and because it is convex it is also strictly quasiconvex.

(iii)

We may note that a function is **concave** if and only if:

$$f[\mu x' + (1-\mu)x''] \geq \mu \cdot f(x') + (1-\mu) \cdot f(x'')$$

where $0 \leq \mu \leq 1$. A function is **convex** if and only if:

$$f[\mu x' + (1-\mu)x''] \leq \mu \cdot f(x') + (1-\mu) \cdot f(x'')$$

where $0 \leq \mu \leq 1$. In this case,

$$\begin{aligned} f[\mu x' + (1-\mu)x''] &= \mu \cdot x' + (1-\mu) \cdot x'' \\ &= \mu \cdot f(x') + (1-\mu) \cdot f(x'') \end{aligned}$$

Thus, this function is both concave and convex, and therefore it is also quasiconcave and quasiconvex. This is true for all linear functions.

(iv)

(a) To test for concavity and convexity, we must form the Hessian matrix of second-order partial derivatives and determine its definiteness. For the problem at hand,

$$\mathbf{H} = \begin{bmatrix} 0 & 1 \\ 1 & 0 \end{bmatrix}$$

Calculating the successive leading principal minors we obtain

$$|\mathbf{H_1}| = 0, |\mathbf{H}| = -1 < 0$$

Thus, the Hessian matrix is indefinite. The function is, therefore, neither convex or concave.

(b) In testing for quasiconcavity and quasiconvexity, we must form the matrix $\mathbf{B}$ which consists of the Hessian matrix with a border of first-order partial derivatives, i.e.,

$$\mathbf{B} = \begin{bmatrix} 0 & \frac{\partial y}{\partial x_1} & \frac{\partial y}{\partial x_2} \\ \frac{\partial y}{\partial x_1} & \frac{\partial^2 y}{\partial x_1^2} & \frac{\partial^2 y}{\partial x_1 \cdot \partial x_2} \\ \frac{\partial y}{\partial x_2} & \frac{\partial^2 y}{\partial x_2 \cdot \partial x_1} & \frac{\partial^2 y}{\partial x_2^2} \end{bmatrix}$$

for the problem at hand we obtain

$$\mathbf{B} = \begin{bmatrix} 0 & x_2 + 1 & x_1 + 1 \\ x_2 + 1 & 0 & 1 \\ x_1 + 1 & 1 & 0 \end{bmatrix}$$

The successive leading principal minors are

$$|\mathbf{D_1}| = \det \begin{bmatrix} 0 & x_2 + 1 \\ x_2 + 1 & 0 \end{bmatrix}$$

which equals $-(x_2 + 1)^2 < 0$.

$$|\mathbf{D_2}| = 0(-1) - (x_2 + 1) \cdot (-(x_1 + 1)) + (x_1 + 1)(x_2 + 1) = 2(x_1 + 1)(x_2 + 1)$$

which is greater than zero, given that $x_1, x_2 \geq 0$. Thus, this function satisfies a sufficient condition for quasiconcavity such that $(-1)^n |\mathbf{D_n}| > 0$ given $x_1, x_2 \geq 0$.

(v)

To test for concavity and convexity, we must form the Hessian matrix of second-order partial derivatives and determine its definiteness. For the problem at hand we obtain

$$\mathbf{H} = \begin{bmatrix} 1 & 0 \\ 0 & 1 \end{bmatrix}$$

the leading principal minors are

$$|\mathbf{H_1}| = 1 > 0, |\mathbf{H}| = 1$$

Thus, the Hessian is positive definite as $|\mathbf{H}_n| > 0$ for all n and for all x which implies that the function is strictly convex. Since the function is strictly convex, it is also strictly quasiconvex.

(vi)

$$\frac{dy}{dx} = e^{-x} + 12 > 0$$

$$\frac{d^2y}{dx^2} = -e^{-x} < 0$$

This satisfies the definition of strict concavity as the second derivative of the function is strictly negative for all x. Because the function is strictly concave it is also strictly quasiconcave. The function also satisfies the sufficient condition for a strictly increasing function such that it is also quasiconvex.

(vii)

Forming the Hessian matrix of second-order partial derivatives we obtain

$$\mathbf{H} = \begin{bmatrix} 6x_1 & 0 & 0 \\ 0 & 6x_2 & 0 \\ 0 & 0 & 6x_3 \end{bmatrix}$$

the leading principal minors are

$$|\mathbf{H}_1| = 6x_1, |\mathbf{H}_2| = 36x_1x_2, |\mathbf{H}| = 216x_1x_2x_3$$

Thus, the Hessian is positive definite for all x if $x_1, x_2, x_3 > 0$. Under this condition, the function is strictly convex which implies it is also strictly quasiconvex.

Homogeneity and Homotheticity

Question 12

(i)

Recall that a function $f(x_1, \ldots, x_n)$ is homogeneous of degree k if

$$f(\lambda x_1, \ldots, \lambda x_n) = \lambda^k \cdot f(x_1, \ldots, x_n),$$

for all $\lambda > 0$. In our case substituting λx for x and λy for y yields the expression

$$\begin{aligned} f(\lambda x, \lambda y) &= A \cdot (\lambda x)^\alpha \cdot \sqrt{(\lambda y)} \\ &= A \cdot \lambda^\alpha \cdot x^\alpha \cdot \sqrt{\lambda} \cdot \sqrt{y} \\ &= \lambda^{\alpha + \frac{1}{2}} \cdot A \cdot x^\alpha \cdot \sqrt{y} \\ &= \lambda^{\alpha + \frac{1}{2}} \cdot f(x, y) \end{aligned}$$

Thus f is homogeneous of degree $\alpha + \frac{1}{2}$.

(ii)

Substituting for x and y, as above, one obtains

$$
\begin{aligned}
f(\lambda x, \lambda y) &= \frac{\lambda x \cdot \lambda y}{\lambda x + \lambda y} \\
&= \frac{\lambda^2 \cdot x \cdot y}{\lambda(x + y)} \\
&= \lambda \cdot \frac{x \cdot y}{(x + y)} = \lambda f(x, y)
\end{aligned}
$$

Thus, the function is homogeneous of degree 1. This is sometimes known as linear homogeneity.

(iii)

Using the same methodology as above:

$$
\begin{aligned}
f(\lambda x, \lambda y) &= \ln(\lambda x + \lambda y) \\
&= \ln(\lambda) + \ln(x + y) \\
&= \ln(\lambda) + f(x, y)
\end{aligned}
$$

In this case the function is not homogeneous, since $f(\lambda x, \lambda y)$ cannot be decomposed into the product of $f(x, y)$ and a power of λ.

(iv)

Proceeding as before,

$$
\begin{aligned}
f(\lambda x, \lambda y) &= \ln\left(\lambda^2 \cdot (x^2 + y^2 + z^2) \right) + 2 \cdot \ln\left(\frac{1}{\lambda} \cdot \frac{2}{x} + \frac{1}{\lambda} \cdot \frac{3}{y} + \frac{1}{\lambda} \cdot \frac{4}{z} \right) \\
&= 2 \cdot \ln(\lambda) + \ln(x^2 + y^2 + z^2) + 2 \cdot \ln\left(\frac{2}{x} + \frac{3}{y} + \frac{4}{z} \right) - 2 \cdot \ln(\lambda) \\
&= f(x, y)
\end{aligned}
$$

Thus the function is homogeneous of degree zero.

Question 13

(i)

Recall that a function is homothetic if it can be written as a strictly monotonic function of a homogenous function. The general procedure to follow is to insert λx and λy into $f()$ in place of x and y, and to see how far we get in finding a homogenous function h. Then we see if the function $f()$ is a strictly monotonic transformation of h.

Following this procedure we find

$$
f(\lambda x, \lambda y) = (\lambda x)^2 (\lambda y)^4 = \lambda^6 x^2 y^4 = \lambda^6 f(x, y)
$$

In this case, the entire function $f(x, y)$ is homogenous of degree 6. Because all homogenous functions are homothetic, the function is homothetic.

(ii)

Substituting λx and λy into f in place of x and y, we find that

$$f(\lambda x, \lambda y) = e^{(\lambda x)^2 + (\lambda y)^2} = e^{(\lambda)^2(x^2 + y^2)}$$

The function $h(x, y) = x^2 + y^2$ is homogeneous of degree 2. However, the whole function $f()$ is not homogenous, because

$$e^{(\lambda)^2(x^2 + y^2)} \neq (\lambda^n) e^{x^2 + y^2}$$

for any value of n.

In order for f to be homothetic, we require the function e^h to be increasing in h. To check whether this is so, we take the deriviative with respect to h:

$$\frac{\partial}{\partial h} e^h = e^h > 0.$$

Because e^h is strictly increasing in h, e^h is therefore a strictly increasing transformation of the homogeneous function h. Thus, $e^{x^2 + y^2}$ is a homothetic function.

(iii) $(x^2 - 1)^2$

The expression x^2 is homogenous of degree 2, but the entire expression is not homogeneous. The question then is whether the function $(h - 1)^2$ is strictly monotonic:

$$\frac{\partial}{\partial h}(h - 1)^2 = 2(h - 1) \not> 0 \text{ for } h \leq 1.$$

The function, therefore, is not monotonic, and so f is not homothetic.

Euler's Theorem

Question 14

The production function is Cobb-Douglas and thus its degree of homogeneity equals the sum of the exponents. The function is, therefore, homogeneous of $\alpha + (1 - \alpha) = 1$.

Euler's Theorem states that for a linearly homogeneous function $f(x_1, x_2)$, it is true that:

$$\frac{\partial f}{\partial x_1} x_1 + \frac{\partial f}{\partial x_2} x_2 \equiv f(x_1, x_2)$$

Thus, for the production function we know that

$$[\alpha K^{\alpha - 1} L^{1 - \alpha}] K + [(1 - \alpha) K^\alpha L^{-\alpha}] L \equiv Q$$
$$\Rightarrow [\alpha K^{\alpha - 1} L^{1 - \alpha}] \frac{K}{L} \equiv \frac{Q}{L} - [(1 - \alpha) K^\alpha L^{-\alpha}]$$

From production theory we note that for the average product of an input to be increasing, its marginal product must be *greater* than its average product. When the average product of labor $(\frac{Q}{L})$ is increasing, therefore, the right hand side of the above identity must be negative because the marginal product of labor $((1 - \alpha) K^\alpha L^{-\alpha})$ exceeds the average product $(\frac{Q}{L})$. Because the left hand side of the identity is equivalent to the right hand side and the quantity of capital (K) can never be less than zero this implies that the marginal product of capital $(\alpha K^{\alpha - 1} L^{1 - \alpha})$ must be decreasing.

Implicit Function Theorem

Question 15

(i)

We can show that this equation does not define x as a function of y. To see this take the square root of both sides: one then has

$$y - x = \pm 1.$$

Thus for any value of x there are two values of y that solve the equation: these are $y = 1 + x$ and $y = x - 1$. Hence y is not a function of x.

(ii)

In this case y is not a function of x: not because there is more than one y for each x, but because there are no (real) values of y and x that solve the equation. In this case we say that the equation is *inconsistent*.

(iii)

In this case the solution is much less obvious, so we will need to appeal to the implicit function theorem. This result is as follows: if at a certain point $(y^0, \mathbf{x}^0)$, the implicit function $F(y, \mathbf{x}) = 0, \partial F/\partial y \neq 0$, and F is continuously differentiable, then y is a continuously differentiable function f of $\mathbf{x}$ in a small area around $(y^0, \mathbf{x}^0)$.

The first step is to check that the equation is not inconsistent, i.e., that there exist values of x and y such that $F(y, \mathbf{x}) = 0$. This is by no means easy, since there is no general method. One approach is take an educated guess: consider those points where $x = y$. For this area, the equation becomes $\ln(2x) = 1$, which is true for $x = e/2$. Thus we know that the equation is consistent for some values of x and y.

Next, we must verify that F is continuously differentiable: in other words we must see if the derivatives are continuous functions. We have

$$\frac{\partial F}{\partial y} = \frac{1}{x + y} + 1, \frac{\partial F}{\partial x} = \frac{1}{x + y} - 1.$$

These expressions exist for all x and y greater than zero, and $\partial F/\partial y$ is never equal to zero. From the theorem we can now deduce that there exists a function $f()$ such that $y = f(x)$, and that $f()$ is continuously differentiable. We cannot write out an algebraic expression for f, but we might be able to do so for the derivative of f.

To obtain this derivative we use the following formula relating total and partial derivatives:

$$\frac{dy}{dx} = -\frac{\partial F/\partial x}{\partial F/\partial y}.$$

This expression comes from the total differential of the equation $F(y, x) = 0$, which is given by

$$\frac{\partial F}{\partial y}dy + \frac{\partial F}{\partial x}dx = 0.$$

Note that we can only use the expression if the implicit function theorem holds. In our case we have

$$\frac{dy}{dx} = -\frac{\frac{1}{x+y} - 1}{\frac{1}{x+y} + 1} = \frac{x+y-1}{x+y+1}$$

This holds for all x and y greater than zero for which $F(y, x) = 0$.

(iv)

Once again we need to apply the implicit function theorem. First we check that there are values of y and x such that the expression holds: we note that indeed the equation is true for $y = x_1 = x_2 = 0$. Next we must check that derivatives exist and are differentiable:

$$\begin{aligned}
\frac{\partial F}{\partial y} &= e^{x_1 + y} + 1 \\
\frac{\partial F}{\partial x_1} &= e^{x_1 + y} \\
\frac{\partial F}{\partial x_2} &= -1
\end{aligned}$$

This proves to be the case. Note also that $\partial F / \partial y$ is not equal to zero: this is another condition that must hold before the theorem can be applied.

We are now in a position to compute the derivatives:

$$\begin{aligned}
\frac{dy}{dx_1} &= -\frac{\partial F / \partial x_1}{\partial F / \partial y} = -\frac{e^{x_1 + y}}{e^{x_1 + y} + 1} \\
\frac{dy}{dx_2} &= -\frac{\partial F / \partial x_2}{\partial F / \partial y} = \frac{1}{e^{x_1 + y} + 1}.
\end{aligned}$$

Question 16

It can be verified that the functions are continuously differentiable. In order to apply the implicit function it remains to check that the Jacobian matrix is non-singular. This matrix is defined as

$$\mathbf{J} = \frac{\partial \mathbf{F}}{\partial \mathbf{y}} = \left(\begin{array}{cc} \partial F_1 / \partial y_1 & \partial F_1 / \partial y_2 \\ \partial F_2 / \partial y_1 & \partial F_2 / \partial y_2 \end{array} \right).$$

For the system under consideration one has

$$\mathbf{J} = \left(\begin{array}{cc} e^{y_1 + x_1} & 1 \\ 0 & 1 \end{array} \right),$$

and so

$$\det(\mathbf{J}) = e^{y_1 + x_1} > 0.$$

With a non-zero determinant of the Jacobian we can now go on to apply the implicit function theorem, which in this case will ensure that there exist continuously differentiable functions f_1 and f_2 such that $y_1 = f_1(\mathbf{x})$ and $y_2 = f_2(\mathbf{x})$. We can also totally differentiate the system of equations $\mathbf{F}(\mathbf{y}, \mathbf{x}) = \mathbf{0}$ to obtain expression for the partial derivatives:

$$\frac{\partial \mathbf{F}}{\partial \mathbf{y}} d\mathbf{y} + \frac{\partial \mathbf{F}}{\partial \mathbf{x}} d\mathbf{x} = \mathbf{0},$$

where

$$\frac{\partial \mathbf{F}}{\partial \mathbf{x}} = \begin{pmatrix} \partial F_1/\partial x_1 & \partial F_1/\partial x_2 \\ \partial F_2/\partial x_1 & \partial F_2/\partial x_2 \end{pmatrix}.$$

Thus, by manipulating the total differential we obtain $\frac{d\mathbf{y}}{d\mathbf{x}} = -\mathbf{J}^{-1}\frac{\partial \mathbf{F}}{\partial \mathbf{x}}$ such that:

$$= -\frac{1}{e^{y_1+x_1}} \begin{pmatrix} 1 & -1 \\ 0 & e^{y_1+x_1} \end{pmatrix} \begin{pmatrix} e^{y_1+x_1} & 1 \\ -1 & 2 \end{pmatrix}$$

$$= -\frac{1}{e^{y_1+x_1}} \begin{pmatrix} e^{y_1+x_1}+1 & -1 \\ -e^{y_1+x_1} & 2e^{y_1+x_1} \end{pmatrix}$$

$$= \begin{pmatrix} -1-[1/e^{y_1+x_1}] & 1/e^{y_1+x_1} \\ 1 & -2 \end{pmatrix}$$

Taylor Series

Question 17

(i)

If f is continuously differentiable $n+1$ times then the Taylor expansion states that

$$f(x) = f(x_0) + f'(x_0)(x-x_0) + \frac{1}{2!}f''(x_0)(x-x_0)^2 + \ldots$$
$$+ \frac{1}{n!}f^n(x_0)(x-x_0)^n + R_{n+1}$$

where R_{n+1} is a remainder term equal to $\frac{1}{n+1!}f^{n+1}(c)(x-x_0)^{n+1}$, c is between x and x_0 and $n!$ is n factorial and signifies $1 \times 2 \times 3 \ldots \times n$. The idea is to choose n large enough so that this remainder term is small. In the present case we set $n = 1$ and the point of approximation is $x_0 = 1$. Thus, the formula becomes

$$\ln(x) = \ln(1) + \frac{1}{1}(x-1) + R_2,$$

where $R_2 = \frac{1}{2}\frac{-1}{c^2}(x-1)^2$.

Given that $\ln(1) = 0$ and $x = 1 + r$ then the first order Taylor series approximation (without the remainder) is, therefore,

$$\ln(1+r) \cong r$$

The difference between the *Taylor expansion* and the *Taylor approximation* is that in the latter case we do not include the remainder term.

(ii)

The first derivative of $f(x)$ is found by using the product rule of differentiation:

$$f'(x) = 2x(x-2)^2 + 2(x-2)x^2$$

The first order approximation around $x_0 = 0$ is, therefore,

$$f(x) \cong 0 + (0)(x) = 0.$$

In fact,

$$f(1) = 1, f(2) = 0.$$

Thus the first order approximation is exact for $x = 2$ but not $x = 1$, despite the fact that $x = 1$ is closer to $x_0 = 0$, the point around which the approximation was made.

Chapter 4

Unconstrained Optimization

Objectives

The questions in this chapter should help the reader solve a variety of unconstrained optimization problems. Readers who can correctly answer all the questions should be able to:

1. Optimize functions of one variable (Questions 1 and 2).

2. Optimize functions of two variables and check the second-order conditions using the Hessian matrix (Questions 3, 4, 5 and 6).

3. Optimize functions of more than two variables and check the second-order conditions using the Hessian matrix (Question 7).

Review

Optimization is a central concept of economics. Agents—be they consumers, workers, or firms—are assumed to maximize or minimize an objective function by choosing the value of variables under their control. For example, a firm's objective function is typically assumed to be its profit function, and the firm is assumed to choose its inputs and output so as to maximize this function. In mathematical economics it is typically assumed that these objective functions are differentiable: when this property holds, the maximum (if it exists) can often be found quickly and easily by using calculus.

A central concept in optimization theory is the *stationary point*. If the function is differentiable, a stationary point is one where the derivative is equal to zero, i.e., $f'(x) = 0$. In the function illustrated below, x_1, x_2 and x_3 are all stationary points. When a function is defined on an open set, such as **R** (set of real numbers), stationarity is a *necessary* condition for a point to be a maximum or a minimum. However, stationarity is not a sufficient condition. The point x_2 in the diagram is not a maximum, rather it is a *point of inflection*.

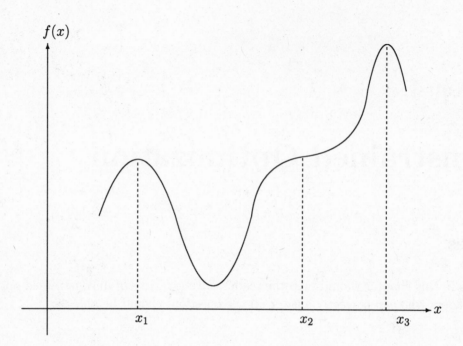

An important distinction must be made between *local optima* and *global optima*. The points x_1 and x_3 both satisfy the criteria for a point x^* to be a *local maximum*, because:

$$f(x^*) \geq f(x) \text{ for all } x \in (x^* - \epsilon, x^* + \epsilon),$$

where $\epsilon > 0$ is as small as necessary. However only x_3 is a *global maximum*, because this is the only point for which:

$$f(x^*) \geq f(x) \text{ for all } x$$

These definitions apply equally to global minima and local minima if the inequalities are reversed.

The point x_2, however, is a stationary point that is neither a local maximum nor a local minimum—instead it is a *point of inflection*. Not all points of inflection are, however, stationary points but a necessary condition for a point of inflection, if the *derivative* function ($f'(x)$) is differentiable, is that $f''(x) = 0$. In this case, the point of inflection represents either a maximum or a minimum of the *derivative* function.

For differentiable functions defined on open sets there is a straightforward method of determining whether the point is a local maximum, a local minimum, or neither. This is often called the *second derivative test*. For functions of one variable, the rule is that a stationary point x^* is a (unique) *local* maximum if;

$$\frac{d^2 f(x^*)}{dx^2} < 0,$$

and a (unique) local minimum if:

$$\frac{d^2 f(x^*)}{dx^2} > 0.$$

If the second derivative of $f(x)$ is zero at x^*, we must make use of the Nth *derivative test* for local extrema. It states that if x^* is a stationary point of an Nth differentiable function $f(x)$ and $f^n(x^*) = \frac{d^n f(x^*)}{dx^n}, n \geq 2$ is the first non-zero derivative of $f(x)$, then

$$f^n(x^*) < 0 \quad \text{and } n \text{ is an even number} \Rightarrow x^* \text{is a local maximum}$$
$$f^n(x^*) > 0 \quad \text{and } n \text{ is an even number} \Rightarrow x^* \text{is a local minimum}$$
$$f^n(x^*) > 0 \quad \text{and } n \text{ is an odd number} \Rightarrow x^* \text{is a point of inflection.}$$

To decide whether or not a stationary point x^* is a **global** optimum we must evaluate the second derivative over the whole domain of the function. If $f''(x) < 0$ for **all** x, then x^* is a *global maximum*. If $f''(x) > 0$ for **all** x, then x^* is a *global minimum*.

For functions of more than one variable, such as

$$y = f(x_1, \ldots, x_n),$$

a stationary point $\mathbf{x}^*$ is one where all the partial derivatives are zero:

$$\frac{\partial f(\mathbf{x}^*)}{\partial x_i} = 0 \; \forall \; i = 1, \ldots, n.$$

To test for local optima we must evaluate the Hessian matrix of second derivatives at $\mathbf{x}^*$. Provided the function is twice continuously differentiable the matrix is defined as

$$\mathbf{H}(\mathbf{x}^*) = \begin{pmatrix} \frac{\partial^2 f(\mathbf{x}^*)}{\partial x_1^2} & \cdots & \frac{\partial^2 f(\mathbf{x}^*)}{\partial x_1 \cdot \partial x_n} \\ \vdots & & \vdots \\ \frac{\partial^2 f(\mathbf{x}^*)}{\partial x_n \cdot \partial x_1} & \cdots & \frac{\partial^2 f(\mathbf{x}^*)}{\partial x_n^2} \end{pmatrix}.$$

If $\mathbf{H}(\mathbf{x}^*)$ is negative definite then $\mathbf{x}^*$ satisfies a sufficient condition to be a local maximum. If $\mathbf{H}(\mathbf{x}^*)$ is positive definite then $\mathbf{x}^*$ satisfies a sufficient condition to be a local minimum. Failure to satisfy these conditions does not necessarily mean that the stationary point is not a maximum or minimum as they are sufficient and not necessary conditions.

The matrix is positive definite if **all** the leading principal minors are positive and is negative definite if $(-1)^n |\mathbf{D_n}| > 0$ where $|\mathbf{D_n}|$ is the nth leading principal minor formed from a matrix of order n (see chapters 2 and 3). If the determinant of the Hessian matrix does **not** equal zero and the leading principal minors do not satisfy the condition for positive or negative definiteness then $\mathbf{x}^*$ satisfies a sufficient condition to be a *saddle-point* which is neither a local maximum nor a minimum. A sufficient condition for x^* to be a unique global optimum is that the Hessian matrix be positive or negative definite over the entire domain of the function.

Further Reading

There are numerous texts that discuss unconstrained maximization and minimization. Useful references on unconstrained optimization in economics include Chiang [13] (chapter 11), Birchenhall and Grout [6] (chapter 5), Baldani et al. [2] (chapters 7 and 8), Hands [22] (chapter 1), Bressler [10] (chapters 9 and 11), Archibald and Lipsey (chapters 6 and 7), Glaister [19] (chapter 15), Silberberg [29] (chapter 6), Rowcroft [26] (chapter 7) and Sydsæter and Hammond [32] (chapters 9 and 17).

Chapter 4 - Questions

Optimization with One Variable

Question 1

(i) Find the unique local maximum of $f(x) = xe^{-x}$.

(ii) Find (if it exists) the maximum of

$$f(x) = \ln(x+1), \text{ where } x > 0.$$

(iii) Find the local maxima of

$$f(x) = -x^2(x-1)(x+1).$$

(iv) Find the local maxima of

$$f(x) = -x^4 + x^3.$$

Question 2

Suppose one has

$$f(x) = \frac{1}{3}x^3 - \alpha x + 1, \text{ where } \alpha \geq 0.$$

For what values of α does $f(x)$ have a unique local maximum, for what values a unique local minimum, and for what values a point of inflection?

Optimization with Two Variables

Question 3

Find the global maximum of

$$f(x, y) = -4x^2 - 4y^2 + 4xy + 2x + y - 5.$$

Question Four

A consumer's utility function is given by

$$U(x, y) = -x^2 - 3y^2 + x + 5y + xy.$$

Find the values of x and y that maximize this function.

Question Five

A firm produces two products, x_1 and x_2, which are priced at p_1 and p_2 respectively. The demand functions for each product are

$$x_1 = D_1(p_1, p_2) = 1 - p_1 + \frac{1}{2}p_2$$

$$x_2 = D_2(p_1, p_2) = 1 + \frac{1}{2}p_1 - p_2.$$

The firm's total cost function is

$$C(x_1, x_2) = x_1 x_2 + x_1 + x_2 + 2.$$

Find the outputs of x_1 and x_2 that maximize the firm's profits, defined as $p_1 x_1 + p_2 x_2 - C(x_1, x_2)$.

Question Six

Suppose that there is a monopolist with a production function

$$y = LK$$

where L is labor and K is capital. The demand curve faced by the firm is assumed to be

$$y = P^{-1/\epsilon}, \frac{1}{2} < \epsilon < 1.$$

The firm can hire as much labor and capital as it wishes at the rates of $w > 0$ and $r > 0$, respectively.

Solve the firm's problem, showing that the values you obtain are a maximum.

Optimization with N Variables

Question Seven

$$\text{Maximize } f(\mathbf{x}) = -\sum_{i=1}^{N}(x_i - b_i)^2,$$

where $b_1, \ldots, b_N$ are constants.

Chapter 4 - Solutions

Optimization with One Variable

Question 1

(i)

The first order condition for this problem is found using the product rule of differentiation

$$f'(x) = e^{-x} + xe^{-x}(-1) = 0$$

Thus we need to find x such that

$$e^{-x} = xe^{-x}$$

Because e^{-x} is never equal to zero we can divide through by this factor to obtain the unique solution

$$x = 1.$$

Thus there is one and only one stationary point of $f(x)$. To verify that it is indeed a unique local maximum we must examine the second derivative:

$$\begin{aligned} f''(x) &= \frac{d}{dx}(e^{-x} - xe^{-x}) \\ &= -e^{-x} - e^{-x} + xe^{-x}) \\ &= -2e^{-x} + xe^{-x} \end{aligned}$$

Thus, $f''(1) = -2e^{-1} + e^{-1} = -e^{-1} < 0$.

Since this expression is strictly negative we have a local maximum at $x = 1$.

(ii)

The first derivative of $f(x)$ is

$$f'(x) = \frac{1}{x+1}.$$

Since this derivative is always positive, there are no stationary points and thus no maximum, despite the fact that the function is concave.

(iii)

Rewritten, the objective function is

$$f(x) = -x^2(x^2 - 1).$$

The first order condition is

$$\begin{aligned} f'(x) &= (-2x)(x^2 - 1) + (-x^2)(2x) \\ &= -2x(2x^2 - 1) = 0 \end{aligned}$$

One stationary value of x is $x = 0$ because any number multiplied by 0 is 0. Other stationary values of x (if they exist) must satisfy

$$2x^2 - 1 = 0.$$

This equation has the solutions $\pm 1/\sqrt{2}$. Thus the stationary values of x are $(-1/\sqrt{2}, 0, 1/\sqrt{2})$. To verify whether any of the solutions is a maximum we must examine the second derivative:

$$
\begin{aligned}
f''(x) &= -2(2x^2 - 1) + -2x(4x) \\
&= -4x^2 + 2 - 8x^2 \\
&= -12x^2 + 2.
\end{aligned}
$$

One then has

$$
\begin{aligned}
f''(-1/\sqrt{2}) &= -12(1/2) + 2 = -4 \\
f''(0) &= 2 \\
f''(1/\sqrt{2}) &= -12(1/2) + 2 = -4
\end{aligned}
$$

Thus, only $\pm 1/\sqrt{2}$ are local maxima: the point zero is a local minimum.

(iv)

Taking the derivative of $f(x)$,

$$f'(x) = -4x^3 + 3x^2 = x^2(3 - 4x),$$

this implies that $x_1 = 0$ and $x_2 = 3/4$ are the stationary points of $f()$. To establish whether these points are local maxima or not we take the second derivative:

$$
\begin{aligned}
f''(x_1) &= -12(x_1)^2 + 6x_1 = 0, \\
f''(x_2) &= -12(x_2)^2 + 6x_2 = -9/4.
\end{aligned}
$$

Thus x_2 is a local maximum, however the results for x_1 are not very informative: thus we must make use of the Nth *derivative test* for local extrema. In our case the third derivative is non-zero:

$$f^3(x_1) = -24x_1 + 6 > 0.$$

Hence x_1 is a point of inflection and so is not a local maximum.

Question 2

The necessary condition for a stationary point is

$$
\begin{aligned}
f'(x) &= x^2 - \alpha = 0 \\
&\Rightarrow x = \sqrt{\alpha}.
\end{aligned}
$$

This implies the stationary points of the function are $\sqrt{\alpha}$ and $-\sqrt{\alpha}$.

Assuming that α is greater than zero, then the second derivative evaluated at the two different points gives

$$f''(\sqrt{\alpha}) = 2\sqrt{\alpha} > 0 \Rightarrow \text{ minimum for all } \alpha > 0$$
$$f''(-\sqrt{\alpha}) = -2\sqrt{\alpha} < 0 \Rightarrow \text{ maximum for all } \alpha > 0$$

Thus, there is a unique local maximum *and* a unique local minimum if α is strictly positive. If, however, $\alpha = 0$ there is just one stationary point, $x = 0$. In this case the second derivative is also zero. The third derivative is

$$f^3(x) = 2.$$

By the Nth derivative test the point $x = 0$ is a point of inflection, and so the function $f()$ has no local maximum or minimum for $\alpha = 0$.

Optimization with Two Variables

Question 3

As in the one variable case, the first step is to find the stationary points. First one takes the partial derivatives:

$$\frac{\partial f}{\partial x} = -8x + 4y + 2 = 0,$$
$$\frac{\partial f}{\partial y} = -8y + 4x + 1 = 0.$$

To solve for x and y, we first equate the two expressions:

$$-8x + 4y + 2 = -8y + 4x + 1$$
$$\Rightarrow -8x - 4x = -8y - 4y - 1$$
$$\Rightarrow -12x = -12y - 1$$
$$\Rightarrow x = y + \frac{1}{12}.$$

Substituting this expression back into the first of the derivatives above we obtain

$$-8(y + \frac{1}{12}) + 4y + 2 = 0$$
$$\Rightarrow -8y - \frac{2}{3} + 4y + 2 = 0$$
$$\Rightarrow -4y = -\frac{4}{3}$$
$$\Rightarrow y = \frac{1}{3}.$$

Consequently,

$$x = \frac{4}{12} + \frac{1}{12}$$
$$\Rightarrow x = \frac{5}{12}.$$

Next we must check the second-order conditions. The Hessian matrix (the matrix of second derivatives) is

$$\mathbf{H} = \begin{pmatrix} \frac{\partial^2 f}{\partial x^2} & \frac{\partial^2 f}{\partial x \partial y} \\ \frac{\partial^2 f}{\partial y \partial x} & \frac{\partial^2 f}{\partial y^2} \end{pmatrix} = \begin{pmatrix} -8 & 4 \\ 4 & -8 \end{pmatrix}$$

The first leading principal minor, $|\mathbf{H}_1|$ is the top left hand element of $\mathbf{H}$, -2. The second principal minor, $|\mathbf{H}_2|$ is the determinant of $\mathbf{H}$. This is equal to $(-8)(-8) - (4)(4) = 48$. Thus, the Hessian is negative definite, the function is strictly concave and the solution is a global maximum.

Question Four

The first step is to find the stationary point(s) of $U(x, y)$, if they exist. Taking the derivative with respect to each variable:

$$\frac{\partial U(x, y)}{\partial x} = -2x + 1 + y$$
$$\frac{\partial U(x, y)}{\partial y} = -6y + 5 + x$$

Setting these two expressions to zero gives us the system of equations

$$y = 2x - 1$$
$$x = 6y - 5,$$

which when solved yields the (unique) stationary point $(x, y) = (1, 1)$.

We must next decide whether or not $(x, y) = (1, 1)$ is a global maximum. This decision requires knowledge of the Hessian matrix of second derivatives, which is given by

$$\mathbf{H} = \begin{pmatrix} -2 & 1 \\ 1 & -6 \end{pmatrix}$$

This matrix is negative definite since the first leading principal minor (-2) is negative and the second $(\det(\mathbf{H}) = 11)$ is positive for any values of x and y. Thus, the objective function is strictly concave and the stationary point $(x, y) = (1, 1)$ is a unique global maximum.

Question Five

Profits, by definition, are equal to total revenue minus total costs. In this case then the profit function can be written as

$$\Pi(p_1, p_2, x_1, x_2) = p_1 x_1 + p_2 x_2 - x_1 x_2 - x_1 - x_2 - 2.$$

The first step in maximizing this function is to try and eliminate p_1 and p_2 using the demand functions. This will reduce the number of variables in the profit function and simplify the calculations.

Rearranging the demand functions we obtain

(1) $p_1 = 1 - x_1 + \frac{1}{2}p_2$
(2) $p_2 = 1 - x_2 + \frac{1}{2}p_1$

Substituting the (2) into (1) we obtain

$$p_1 = 1 - x_1 + \frac{1}{2} \cdot (1 - x_2 + \frac{1}{2}p_1)$$
$$= \frac{3}{2} - x_1 - \frac{1}{2}x_2 + \frac{1}{4}p_1,$$
$$= \frac{4}{3} \cdot (\frac{3}{2} - x_1 - \frac{1}{2}x_2),$$
$$= 2 - \frac{4}{3}x_1 - \frac{2}{3}x_2.$$

Inserting this expression into (2) implies

$$p_2 = 1 - x_2 + \frac{1}{2}(2 - \frac{4}{3}x_1 - \frac{2}{3}x_2),$$
$$= 1 - x_2 + 1 - \frac{2}{3}x_1 - \frac{1}{3}x_2$$
$$= 2 - \frac{2}{3}x_1 - \frac{4}{3}x_2.$$

These expressions can now be substituted back into the profit function:

$$\Pi(x_1, x_2) = (2 - \frac{4}{3}x_1 - \frac{2}{3}x_2) \cdot x_1 + (2 - \frac{2}{3}x_1 - \frac{4}{3}x_2) \cdot x_2$$
$$- x_1 x_2 - x_1 - x_2 - 2$$
$$= x_1 + x_2 - \frac{4}{3}x_1^2 - \frac{4}{3}x_2^2 - \frac{7}{3}x_1 x_2 - 2$$

The next step is to find the stationary point(s) of this function. The necessary conditions for a maximum are:

$$\frac{\partial \Pi(x_1, x_2)}{\partial x_1} = 1 - \frac{8}{3}x_1 - \frac{7}{3}x_2 = 0$$
$$\frac{\partial \Pi(x_1, x_2)}{\partial x_2} = 1 - \frac{8}{3}x_2 - \frac{7}{3}x_1 = 0$$

Manipulation of the necessary conditions reveals

$$\frac{8}{3}x_1 + \frac{7}{3}x_2 = \frac{8}{3}x_2 + \frac{7}{3}x_1$$
$$\Rightarrow x_1 = x_2.$$

Substituting $x_1 = x_2 = x$ into $\frac{\partial \pi(x_1, x_2)}{\partial x_1}$ we obtain

$$1 - \frac{8}{3}x - \frac{7}{3}x = 0$$
$$1 = \frac{15}{3}x$$
$$\Rightarrow x = \frac{3}{15} = \frac{1}{5}$$

The next step is to examine the second-order conditions. The Hessian matrix is

$$\mathbf{H} = \begin{pmatrix} -\frac{8}{3} & -\frac{7}{3} \\ -\frac{7}{3} & -\frac{8}{3} \end{pmatrix}$$

This matrix is negative definite since $|\mathbf{H}_1| = -\frac{8}{3} < 0$ and

$$|\mathbf{H}_2| = (-\frac{8}{3})^2 - (-\frac{7}{3})^2 = \frac{5}{3} > 0$$

Thus, the stationary point $(\frac{1}{5}, \frac{1}{5})$ is a local maximum of the profit function.

Question Six

The firm's problem is to

$$\text{Maximize } \Pi(L, K, y, P) = Py - wL - rK.$$

Inverting our expression for y gives an expression for P:

$$P = y^{-\epsilon}, \frac{1}{2} < \epsilon < 1.$$

Substituting for $y = L \cdot K$ allows us to rewrite the profit function as

$$\Pi(L, K) = (L \cdot K)^{1-\epsilon} - wL - rK$$

The first order conditions are

$$\Pi_L(L, K) = (1 - \epsilon)K(KL)^{-\epsilon} - w = 0$$
$$\Pi_K(L, K) = (1 - \epsilon)L(KL)^{-\epsilon} - r = 0$$

The usual approach to solving first-order conditions of this kind is to "divide" one equation by the other so that we can cancel out the common expressions:

$$\frac{(1 - \epsilon) \cdot K \cdot (K \cdot L)^{-\epsilon}}{(1 - \epsilon) \cdot L \cdot (K \cdot L)^{-\epsilon}} = \frac{w}{r}.$$

(We can only do this because $r > 0$ implies that $(1 - \epsilon) \cdot L \cdot (K \cdot L)^{1-\epsilon}$ is not 0). Thus, we obtain the expression

$$\frac{K}{L} = \frac{w}{r} \Rightarrow K = \frac{w \cdot L}{r}.$$

Substituting back into the equation $\Pi_L = 0$ we obtain

$$(1 - \epsilon) \cdot \frac{w}{r} \cdot L \cdot (\frac{w}{r} \cdot L^2)^{-\epsilon} = w$$

which implies that

$$(1 - \epsilon) \cdot \frac{1}{r} \cdot \left(\frac{w}{r} \right)^{-\epsilon} \cdot L^{1-2\epsilon} = 1.$$

We want to solve for L. The best approach is to put the terms in L on one side and everything else on the other:

$$L^{1-2\epsilon} = \frac{1}{1 - \epsilon} \cdot r \cdot r^{-\epsilon} \cdot w^{\epsilon}$$
$$\Rightarrow L^{1-2\epsilon} = \frac{1}{1 - \epsilon} \cdot r^{1-\epsilon} \cdot w^{\epsilon}$$
$$\Rightarrow L^* = \left(\tfrac{1}{1-\epsilon} \cdot r^{1-\epsilon} \cdot w^{\epsilon} \right)^{1/(1-2\epsilon)}$$

The last step is only possible because $\epsilon > \frac{1}{2}$ implies that $(1 - 2\epsilon) < 0$.

We can also obtain a similar expression for K.

$$K^* = (\frac{1}{1 - \epsilon} \cdot w^{1-\epsilon} \cdot r^\epsilon)^{1/(1-2\epsilon)}$$

Are these solutions a maximum? To check this we go back to the profit function and check whether or not it is concave. One has

$$\Pi_{LL} = -\epsilon \cdot (1 - \epsilon) \cdot K^2 \cdot (K \cdot L)^{-\epsilon-1} < 0$$
$$\Pi_{KL} = \Pi_{LK} = (1 - \epsilon)^2 \cdot (K \cdot L)^{-\epsilon} > 0$$
$$\Pi_{KK} = -\epsilon \cdot (1 - \epsilon) \cdot L^2 \cdot (K \cdot L)^{-\epsilon-1} < 0.$$

In consequence the Hessian matrix of Π is

$$\mathbf{H} = (1 - \epsilon)(K \cdot L)^{-\epsilon} \begin{pmatrix} -\epsilon(K \cdot L)^{-1}K^2 & (1 - \epsilon) \\ (1 - \epsilon) & -\epsilon(K \cdot L)^{-1}L^2 \end{pmatrix}$$

or

$$\mathbf{H} = (1 - \epsilon)(K \cdot L)^{-\epsilon} \begin{pmatrix} -\epsilon(K/L) & (1 - \epsilon) \\ (1 - \epsilon) & -\epsilon(L/K) \end{pmatrix}$$

Thus, sign $|\mathbf{H_1}| = \text{sign}[-\epsilon(K/L)] < 0$, and:

$$\begin{aligned} \text{sign}(\det(\mathbf{H})) &= \text{sign}[\epsilon^2(K/L)(L/K) - (1 - \epsilon)^2] \\ &= \text{sign}[\epsilon^2 - (1 - \epsilon)^2] \\ &= \text{sign}[-1 + 2\epsilon] > 0 \text{ since } \epsilon > \frac{1}{2} \end{aligned}$$

Thus, our Hessian is negative definite and the profit function is concave for all values of L and K. It follows that the stationary point is a global maximum.

Optimization with N Variables

Question Seven

There are N first order conditions:

$$\frac{\partial f}{\partial x_1} = -2(x_1 - b_1) = 0$$
$$\frac{\partial f}{\partial x_2} = -2(x_2 - b_2) = 0$$
$$\frac{\partial f}{\partial x_N} = -2(x_N - b_N) = 0$$

Thus, the stationary point is $\mathbf{x} = (b_1, \ldots, b_N)$. Is this point a maximum? The Hessian is

$$\mathbf{H} = \begin{pmatrix} -2 & 0 & \cdots & 0 \\ 0 & -2 & \cdots & 0 \\ \vdots & \vdots & & \vdots \\ \vdots & \vdots & & \vdots \\ 0 & 0 & \cdots & -2 \end{pmatrix}$$

The leading principal minors are

$$
\begin{aligned}
|\mathbf{H_1}| &= -2 < 0 \\
|\mathbf{H_2}| &= 4 > 0
\end{aligned}
$$

.   .

.   .

It follows that,

$$
|\mathbf{H_N}| = (-2)^N = \begin{cases} > 0 \text{ if N is even} \\ < 0 \text{ if N is odd} \end{cases}
$$

Thus, the Hessian is negative definite and so the function is strictly concave and our solution is a unique global maximum.

Chapter 5

Optimization with Equality Constraints

Objectives

The questions in this chapter give the reader the opportunity to solve a variety of constrained optimization problems. Readers who can correctly answer all the questions should be able to:

1. Solve constrained optimization problems with two variables and one linear constraint and check the second-order conditions (Questions 1 and 2).

2. Solve two variable and one nonlinear constraint optimization problems (Questions 3, 4, 5 and 6).

3. Solve a multiple variable optimization problem with two nonlinear constraints (Question 7).

Review

In this chapter we study *constrained* optimization. This means that, unlike in the previous chapter, we are not free to choose any value of the choice variables. Instead, the solutions that are selected must conform to at least one constraint. For example, when a consumer attempts to maximize her utility, typically she is constrained by the requirement that her total purchases cannot exceed her income.

A special type of constraint is the equality constraint, which takes the form $g(\mathbf{x}) = 0$, where g is the constraint function and $\mathbf{x} = (x_1, x_2, \dots, x_n)$ is the vector of choice variables. For example, if a consumer is obliged to spend **all** of her income, then the constraint is an equality constraint: purchases must not be greater or less than her income.

A more general kind of constraint is the inequality constraint. For example, we might assume that the choice variables are constrained to be greater or equal to zero. The mathematics required to handle this kind of problem are more advanced than those required to deal with equality constraints, and so we postpone discussion of inequality constraints until chapters 7 and 8.

In a typical equality constrained optimization problem we maximize the objective function where the choice variables are defined by the vector $\mathbf{x}$.

$$\max_{\mathbf{x}} f(\mathbf{x})$$

subject to the constraints:

$$
\begin{aligned}
g_1(\mathbf{x}) &= c_1 \\
g_2(\mathbf{x}) &= c_2 \\
&\vdots \\
g_m(\mathbf{x}) &= c_m
\end{aligned}
$$

The principal method of solving constrained optimization problems with equality constraints is to use *the method of Lagrange multipliers*. Heuristically, this procedure can be divided into three stages:

1. Transform the original maximization problem into a different problem involving *Lagrange multipliers*.

2. Find the stationary point(s) for the transformed problem by solving the first-order conditions. The Lagrange Theorem implies, under certain conditions, that any stationary point—$\mathbf{x}_0$—of the transformed problem is a stationary point of the original problem. This does not mean that a stationary point is a solution to the original problem, but it does mean that **if** there is a solution, it will be a stationary point.

3. Check the stationary points using the second-order conditions to ensure that any stationary point maximizes the objective function given the constraints.

As with unconstrained optimization, the first-order conditions are *necessary* conditions: any candidate for a solution must satisfy these conditions, but they are not enough by themselves to indicate whether a solution has been found. For this we need to refer to the second-order conditions. If the first-order conditions are satisfied, the second-order conditions are *sufficient* to guarantee that a solution has been found.

We will now go through the three stages of the method of Lagrange multipliers in greater detail.

To solve constrained optimization problems we first write down the *Lagrangean function*, which takes the following form:

$$
L(\mathbf{x}, \lambda_1, \ldots \lambda_m) = f(\mathbf{x}) + \lambda_1(c_1 - g_1(\mathbf{x})) + \lambda_2(c_2 - g_2(\mathbf{x})) + \ldots + \lambda_m(c_m - g_m(\mathbf{x})).
$$

The symbols λ_i denote *Lagrange multipliers*, which are treated as choice variables, like $\mathbf{x}$, rather than as parameters. The Lagrange multiplier λ_i for the ith constraint can be interpreted as a *shadow price* of the resource i, or the rate at which the optimal value of the objective function changes with respect to a change in c_i.

In order to use the Lagrange multiplier method, it must be true that the number of choice variables (n) in the original objective function exceeds the number of independent constraints (m). The theorem also requires $f(\mathbf{x})$ and $g(\mathbf{x})$ to have continuous partial derivatives.

The next step is to solve for the stationary points of the Lagrangean function defined by $\mathbf{x}_0$, and the associated Lagrange multipliers λ_0. The method of solution requires us to partially differentiate the Lagrangean function with respect to each $x_1, x_2, \ldots x_n$ and each $\lambda_1, \lambda_2, \ldots \lambda_m$ setting each partial derivative equal to zero. Using these first-order conditions we solve for the values of $\mathbf{x}$ and λ that satisfy the equations. In some problems there may be multiple stationary points and associated Lagrangean multipliers.

The third step is to determine whether $\mathbf{x}_0$ is a constrained maximum of the function $f(\mathbf{x})$. If the Lagrangean function is globally concave, the stationary point is a maximum. If it is convex, the stationary point is a minimum.

Often, however, the function will not have global curvature properties that can be readily identified, and so we have to resort to second-order sufficient conditions for *local* maxima. It may be that there are several local maxima, in which case we have to check which is the global maximum by substituting the different maxima into the objective function $f(\mathbf{x})$. It is important to remember that the second-order conditions are *sufficient*, not *necessary* conditions. Failure to satisfy them does not necessarily mean that the stationary point is not a maximum (or minimum).

We turn now to these second-order conditions for a local maximum or minimum. These sufficient conditions are obtained from the sign of the leading bordered principal minors (see chapters 2 and 3) of the bordered Hessian matrix, which consists of an *interior* matrix of second-order partial derivatives with a border for each constraint. This bordered Hessian can be written as

$$\overline{\mathbf{H}} = \begin{pmatrix} 0 & \cdots & 0 & -\frac{\partial g_1}{\partial x_1} & \cdots & -\frac{\partial g_1}{\partial x_n} \\ \vdots & & \vdots & \vdots & & \vdots \\ 0 & \cdots & 0 & -\frac{\partial g_m}{\partial x_1} & \cdots & -\frac{\partial g_m}{\partial x_n} \\ -\frac{\partial g_1}{\partial x_1} & \cdots & -\frac{\partial g_m}{\partial x_1} & \frac{\partial^2 L}{\partial x_1^2} & \cdots & \frac{\partial^2 L}{\partial x_1 \partial x_n} \\ \vdots & & \vdots & \vdots & & \vdots \\ -\frac{\partial g_1}{\partial x_n} & \cdots & -\frac{\partial g_m}{\partial x_n} & \frac{\partial^2 L}{\partial x_n \partial x_1} & \cdots & \frac{\partial^2 L}{\partial x_n^2} \end{pmatrix}$$

The sufficient conditions for local extrema require us to examine the determinants of the leading principal minors of $\overline{\mathbf{H}}$. Recall from chapter 2 that the leading principal minors comprise the element in the top left hand corner of the matrix, the 2×2 matrix in the top left hand corner, the 3×3 matrix in the top left hand corner, etcetera, up to and including the matrix itself. However, in this case, we are only interested in the last $n - m$ determinants of the $n + m$ leading principal minors, $|\overline{\mathbf{H}}_{2m+1}|, |\overline{\mathbf{H}}_{2m+2}|, \ldots |\overline{\mathbf{H}}_{m+n}|$, where $|\overline{\mathbf{H}}_{m+n}|$ is the determinant of the bordered Hessian itself. Where the number of constraints (m) is one fewer than the number of choice variables (n) then $2m + 1 = n + m$ and we need only examine the determinant of the bordered Hessian itself.

A sufficient condition for a local maximum is that the determinant of each of the last $n - m$ leading principal minors, $\det(\overline{\mathbf{H}}_r)$, where $r = m+1, m+2, \ldots, n$ and is the dimension of the *interior* (not including the border) of the bordered Hessian matrix multiplied by $(-1)^r$, is strictly positive, i.e., $(-1)^r \det(\overline{\mathbf{H}}_r) > 0$. A sufficient condition for a local minimum is that the determinant of each of the last $n - m$ leading principal minors, $\det(\overline{\mathbf{H}}_r)$, where $r = m + 1, m + 2, \ldots, n$, and is the dimension of the interior (not including the border) of the bordered Hessian matrix multiplied by $(-1)^m$, is strictly positive, i.e., $(-1)^m \det(\overline{\mathbf{H}}_r) > 0$.

Further Reading

As in the previous chapter good references for constrained optimization with equality constraints are Chiang [13](chapter 12), Glaister [19] (chapter 17), Birchenall and Grout [6] (chapter 8), Baldani et al. [2] (chapters 9 and 10), Archibald and Lipsey [1] (chapters 10 and 11), Rowcroft [26] (chapter 19.3), Silberberg [29] (chapter 6) and Sydsæter and Hammond [32] (chapter 18). We highly recommend Sydsæter and Hammond [32] for a thorough but very readable discussion of the method of Lagrange. A concise exposition can be found in the mathematical appendix of Varian [35]. A more advanced reference is Sydsæter [31] (chapters 5, parts 9-11).

Chapter 5 - Questions

Optimization with One Linear Constraint

Question 1

Suppose that a consumer has an objective function of the form

$$U(x_1, x_2) = \ln(x_1 + 1) + \ln(x_2 + 2)$$

where all the parameters are strictly positive. Furthermore, there is a budget constraint which requires that

$$2x_1 + x_2 = M,$$

where M is the consumer's income. By maximizing the objective function subject to the constraint, find the consumer's demand functions: that is, obtain expressions of the form $x_i = f_i(M)$ for $i = 1, 2$.

Question 2

Minimize:

$$y = x_1^2 + x_2^2$$

subject to:

$$x_1 + x_2 = 1$$

Optimization with One Nonlinear Constraint

Question 3

Maximize:

$$y = x_1 + x_2$$

subject to:

$$\frac{1}{2} = x_1^2 + x_2^2.$$

Question 4

Suppose that a firm uses inputs x_1 and x_2 to produce an output $\bar{q}$ according to the production function

$$\bar{q} = (x_1 \cdot x_2)^2,$$

where the level of $\bar{q}$ is determined exogenously. The firm must pay a price of w_1 for each unit of x_1 and w_2 for each unit of x_2 it uses. Solve the firm's cost minimizing problem.

Question 5

Find the values for x_1 and x_2 that solve the following problem:

Maximize:

$$y = 3x_1^2 + 2x_2^2 - 4x_2 + 1$$

subject to:

$$x_1^2 + x_2^2 = 16$$

Question 6

Find the stationary point of the function:

$$y = cx_1 + x_2$$

subject to the constraint:

$$x_1 + \ln(x_2) = b$$

where b and c are both greater than zero.

Determine through the second-order conditions whether the stationary point is a maximum or a minimum.

Optimization with Multiple Constraints

Question 7

In this problem we move from objective functions of two variables to objective functions of three variables. The problem is to maximize

$$y = x_1 + x_2 + x_3$$

subject to:

$$x_1^2 + x_2^2 = 1$$
$$x_2^2 + x_3^2 = 1$$

(The constraints can be pictured as open ended cylinders of infinite length.)

Chapter 5 - Solutions

Optimization with One Linear Constraint

Question 1

One way of solving this problem would be to solve the constraint for x_1 as a function of x_2 and then to substitute the resulting expression back into the objective function. However the method, although feasible in the present case, is not generally applicable since it may not always be possible to solve the contraint function in terms of one variable as a unique function of the other variables. Instead, we shall use the method of Lagrange multipliers outlined in the introduction to this chapter.

The Lagrangean function is:

$$L(x_1, x_2, \lambda) = \ln(x_1 + 1) + \ln(x_2 + 2) + \lambda(M - 2x_1 - x_2)$$

In order to maximize L we must obtain the first-order conditions:

$$\frac{\partial L}{\partial x_1} = \frac{1}{x_1 + 1} - 2\lambda = 0$$

$$\frac{\partial L}{\partial x_2} = \frac{1}{x_2 + 2} - \lambda = 0,$$

$$\frac{\partial L}{\partial \lambda} = M - 2x_1 - x_2 = 0.$$

Rearranging slightly the three first-order conditions: become

$$\frac{1}{2} \frac{1}{x_1 + 1} = \lambda,$$

$$\frac{1}{x_2 + 2} = \lambda,$$

$$2x_1 + x_2 = M$$

The first step in this type of problem is usually to eliminate λ. We have made this easy for ourselves in the present case: the first two equations are both equal to λ, thus we can set the left hand sides equal to each other and solve for x_1 as a function of x_2:

$$\frac{1}{2} \frac{1}{x_1 + 1} = \frac{1}{x_2 + 2}$$

$$\Rightarrow x_1 = (x_2 + 2)\frac{1}{2} - 1 = \frac{x_2}{2}.$$

Having gone as far as we can go with these two equations we turn our attention to the budget constraint. Rewriting this as a function of x_1 and substituting into the expression above we obtain:

$$x_1 = \frac{M - 2x_1}{2} = \frac{M}{2} - x_1$$

$$\Rightarrow x_1 = \frac{M}{4}$$

Thus

$$x_2 = 2x_1 = x_1 = \frac{M}{2}.$$

As with an unconstrained problem, there is no guarantee that a stationary point is necessarily a maximimum. However, because the objective function is concave, and the constraints are linear (and thus concave and convex) the Lagrangean function is concave. It follows that the stationary point we have found is indeed a global maximum. The graphical solution is illustrated below.

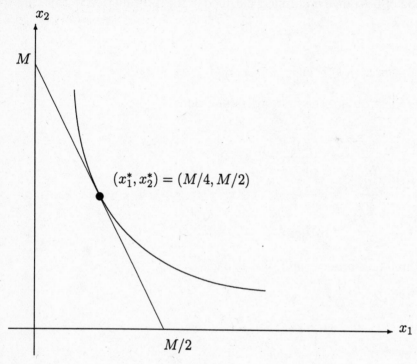

Question 2

Minimizing a function is the same as maximizing the negative of the function. Forming the Lagrangean of the equivalent maximization problem yields:

$$L(x_1, x_2, \lambda) = -x_1^2 - x_2^2 + \lambda(1 - x_1 - x_2)$$

The necessary conditions for a maximum are:

(1) $\frac{\partial L}{\partial x_1} = -2x_1 - \lambda = 0$

(2) $\frac{\partial L}{\partial x_2} = -2x_2 - \lambda = 0$

(3) $\frac{\partial L}{\partial \lambda} = 1 - x_1 - x_2 = 0$

From (1) and (2), we have: $-2x_1 = -2x_2$ or $x_1 = x_2$ Substituting this result into (3) yields: $1 - 2x_1 = 0$. Thus,

$$x_1 = \frac{1}{2} \text{ and } x_2 = \frac{1}{2}.$$

This implies that from (1), $\lambda = 1$.

The objective function is concave and the constraints are linear (and thus concave and convex). It follows, therefore, that the Lagrangean function is concave, and so the solution is a global maximum.

Alternatively, we could have checked for a local optimum using the bordered Hessian

$$\overline{\mathbf{H}} = \begin{pmatrix} 0 & -1 & -1 \\ -1 & -2 & 0 \\ -1 & 0 & -2 \end{pmatrix}.$$

Because we have one fewer constraint than choice variables, we need only examine the determinant of the entire matrix. Expanding along the first row we find that

$$\det(\overline{\mathbf{H}}) = (-1)^3(-1)(2) + (-1)^4(-1)(-2) = 4 > 0.$$

Thus in this case $r = m + 1 = 2$ and $(-1)^2 \det(\overline{\mathbf{H}}) > 0$ and the solution is a local maximum.

The question could also have been solved as a minimization problem in which case $\det(\overline{\mathbf{H}}) = -4$ and $(-1)^1 \det(\overline{\mathbf{H}}) > 0$ which satisfies the sufficient condition for a minimum. The solution to the minimization problem is illustrated below.

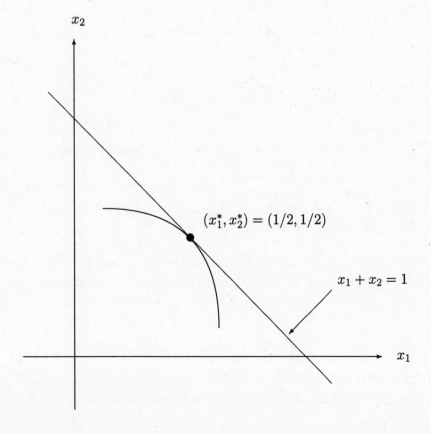

Optimization with One Nonlinear Constraint

Question 3

The Lagrangean function is given by:

$$L(x_1, x_2, \lambda) = x_1 + x_2 + \lambda(\frac{1}{2} - x_1^2 - x_2^2)$$

The first-order necessary conditions are:
(1) $\frac{\partial L}{\partial x_1} = 1 + \lambda(-2x_1) = 0$
(2) $\frac{\partial L}{\partial x_2} = 1 + \lambda(-2x_2) = 0$
(3) $\frac{\partial L}{\partial \lambda} = \frac{1}{2} - x_1^2 - x_2^2 = 0$

Equating the left hand sides of equations (1) and (2) gives

$$\lambda(-2x_1) = \lambda(-2x_2).$$

Cancelling -2λ from both sides implies that $x_1 = x_2$. Note that dividing by -2λ is only permissible if $\lambda \neq 0$. This condition is satisfied in the present case because if λ were equal to zero then equation (1) would imply that $1 = 0$, which is false.

Inserting the equality $x_1 = x_2$ into (3) produces:

$$\frac{1}{2} = 2x_1^2.$$

This equation has the solutions $\pm\frac{1}{2}$, which implies that there are two solutions to the first-order conditions:

$$(x_1, x_2, \lambda) = (0.5, 0.5, 1) \quad \text{and} \quad (x_1, x_2, \lambda) = (-0.5, -0.5, -1).$$

These solutions give the values of 1 and -1 for y.

The next step is to check the second-order conditions. The bordered Hessian is:

$$\overline{\mathbf{H}} = \begin{pmatrix} 0 & -2x_1 & -2x_2 \\ -2x_1 & -2\lambda & 0 \\ -2x_2 & 0 & -2\lambda \end{pmatrix}.$$

Expanding along the first row we can determine if $|\overline{\mathbf{H}}| > 0$. If this is the case, we have found a maximum because $r = m + 1 = 2$ and $m = 1$ and a sufficient condition for a local maximum is $(-1)^r \det \overline{\mathbf{H_r}} > 0$.

$$\det(\overline{\mathbf{H}}) = (-1)^3(-2x_1) \cdot (4x_1\lambda) + (-1)^4(-2x_2)(-4x_2\lambda) = \lambda(8x_1^2 + 8x_2^2)$$

Thus, when $(x_1, x_2, \lambda) = (0.5, 0.5, 1)$, the $\det(\overline{\mathbf{H}}) > 0$, and we have found a local maximum. When $(x_1, x_2, \lambda) = (-0.5, -0.5, -1)$, the $\det(\overline{\mathbf{H}}) < 0$, and we have found a local minimum because $(-1)^m \det \overline{\mathbf{H_r}} > 0$.

Question 4

The firm's problem is to minimize costs given the two constraints. Formally this is written as

$$\min_{x_1, x_2} C = w_1 x_1 + w_2 x_2$$

subject to:

$$\overline{q} = (x_1 \cdot x_2)^2$$

Rewriting this in the form of maximization problem one obtains the Lagrangean:

$$L(x_1, x_2, \lambda) = -w_1 x_1 - w_2 x_2 + \lambda \cdot \left(-\overline{q} + (x_1 \cdot x_2)^2 \right).$$

The first-order conditions are

(1) $\frac{\partial L}{\partial x_1} = -w_1 + \lambda \cdot 2x_1 \cdot x_2^2 = 0$

(2) $\frac{\partial L}{\partial x_2} = -w_2 + \lambda \cdot 2x_2 \cdot x_1^2 = 0$

(3) $\frac{\partial L}{\partial \lambda} = (x_1 \cdot x_2)^2 - \overline{q} = 0$

The first task is to eliminate λ. This can be done by using equations (1). Rearranging one has

$$\lambda = \frac{w_1}{2x_1 \cdot x_2^2}.$$

Substituting this equation into (2) gives

$$w_2 = \frac{w_1}{2x_1 \cdot x_2^2} \cdot 2x_2 \cdot x_1^2$$

$$\Rightarrow w_2 = \frac{w_1 x_1}{x_2}$$

$$\Rightarrow x_1 = \frac{w_2}{w_1} \cdot x_2.$$

Using (3),

$$\left(\frac{w_2}{w_1} \cdot x_2 \cdot x_2\right)^2 = \overline{q}$$

$$\Rightarrow x_2 = \overline{q}^{\frac{1}{4}} \cdot \left(\frac{w_1}{w_2} \right)^{\frac{1}{2}},$$

$$\Rightarrow x_1 = \overline{q}^{\frac{1}{4}} \cdot \left(\frac{w_2}{w_1} \right)^{\frac{1}{2}}.$$

The next step is to check the second-order conditions using the bordered Hessian.

$$\overline{\mathbf{H}} = \begin{pmatrix} 0 & -2x_1 x_2^2 & -2x_2 x_1^2 \\ -2x_1 x_2^2 & 2\lambda x_2^2 & 4\lambda x_1 x_2 \\ -2x_2 x_1^2 & 4\lambda x_1 x_2 & 2\lambda x_1^2 \end{pmatrix}.$$

Expanding along the first row we can determine if $\det(\overline{\mathbf{H}}) > 0$. If this is the case we have a sufficient condition for a maximum (remembering that we're solving the question as a maximization problem) because $(-1)^r \det \overline{\mathbf{H}}_r > 0$ where $r = m + 1 = 2$.

$$\det(\overline{\mathbf{H}}) = (-1)^3 (2x_1 x_2^2) \cdot (4\lambda x_1^3 x_2^2 - 8\lambda x_1^3 x_2^2) + (-1)^4 (2x_2 x_1^2)(8\lambda x_1^2 x_2^3 - 4\lambda x_1^2 x_2^3) = 16\lambda x_1^4 x_2^4 > 0$$

Question 5

Forming the Lagrangean yields:

$$L(x_1, x_2, \lambda) = 3x_1^2 + 2x_2^2 - 4x_2 + 1 + \lambda(16 - x_1^2 - x_2^2)$$

The necessary conditions for the maximum are:

(1) $\frac{\partial L}{\partial x_1} = 6x_1 - 2x_1 \lambda = 0$
(2) $\frac{\partial L}{\partial x_2} = 4x_2 - 4 - 2x_2 \lambda = 0$
(3) $\frac{\partial L}{\partial \lambda} = 16 - x_1^2 - x_2^2 = 0$

Using only equation (1), we note that $\frac{6x_1}{2x_1} = \lambda = 3$.

Substitute $\lambda = 3$ into equation (2):

$$4x_2 - 4 - 6x_2 = 0$$

Thus: $-2x_2 = 4$ and so $x_2 = -2$.
Substituting $x_2 = -2$ into (3) yields $x_1^2 = 12$.
Thus $x_1 = \pm\sqrt{12}$.

The two complete solutions are:

$$(x_1, x_2, \lambda, y) = (\sqrt{12}, -2, 3, 53)$$

$$(x_1, x_2, \lambda, y) = (-\sqrt{12}, -2, 3, 53)$$

Another way to approach this question is as follows,

Note that $x_1 = 0$ satisfies (1). If $x_1 = 0$, then from (3) we have $x_2 = \pm 4$.

If $x_2 = +4$, then from (2): $4 \cdot 4 - 4 - 2 \cdot \lambda \cdot 4 = 0$. Thus, $12 = 8\lambda$ and $\lambda = 1.5$.

If $x_2 = -4$, then from (2): $4 \cdot (-4) - 4 - 2 \cdot \lambda \cdot (-4) = 0$. Thus, $-20 = -8\lambda$ and $\lambda = 2.5$.

Thus, we have two other possible solutions that satisfy the first order conditions.

$$(x_1, x_2, \lambda, y) = (0, 4, 1.5, 17)$$

$$(x_1, x_2, \lambda, y) = (0, -4, 2.5, 49)$$

We can determine the values of x_1 and x_2 that provide a maximum by finding the y value for each of the stationary points. Alternatively, we can check the second-order conditions using the bordered Hessian. The bordered Hessian in this case is

$$\begin{bmatrix} 0 & -2x_1 & -2x_2 \\ -2x_1 & 6 - 2\lambda & 0 \\ -2x_2 & 0 & 4 - 2\lambda \end{bmatrix}$$

There are four cases to be considered. Expanding by row 1 we can calculate the cofactors and the determinant:

(1) If $(x_1, x_2, \lambda, y) = (\sqrt{12}, -2, 3, 53)$, then:

$$\det(\overline{\mathbf{H}}) = 0 \cdot 0 + 2\sqrt{12} \cdot 4\sqrt{12} + 4 \cdot 0 = 96 > 0$$

Thus, this solution is a maximum because $r = 2$ and $(-1)^2 \det(\overline{\mathbf{H}}) > 0$.

(2) If $(x_1, x_2, \lambda, y) = (-\sqrt{12}, -2, 3, 53)$, then:

$$\det(\overline{\mathbf{H}}) = 0 \cdot 0 - 2\sqrt{12} \cdot -4\sqrt{12} + 4 \cdot 0 = 96 > 0$$

Thus, this solution is a maximum because $r = 2$ and $(-1)^2 \det(\overline{\mathbf{H}}) > 0$.

(3) If $(x_1, x_2, \lambda, y) = (0, 4, 1.5, 17)$, then:

$$\det(\overline{\mathbf{H}}) = 0 \cdot 15 - 0 \cdot 0 - 8 \cdot 24 = -192 < 0$$

Thus, this solution is a minimum because $m = 1$ and $r = 2$ and $(-1)^1 \det(\overline{\mathbf{H}}) > 0$.

(4) If $(x_1, x_2, \lambda, y) = (0, -4, 2.5, 49)$, then:

$$\det(\overline{\mathbf{H}}) = 0 \cdot (-1) - 0 \cdot 0 + 8 \cdot (-8) = -64 < 0$$

Thus, this solution is a minimum because $m = 1$ and $r = 2$ and $(-1)^1 \det(\overline{\mathbf{H}}) > 0$.

Question 6

Forming the Lagrangean yields:

$$L(x_1, x_2, \lambda) = cx_1 + x_2 + \lambda(b - x_1 - \ln(x_2))$$

The necessary conditions for a stationary point are:

(1) $\quad \frac{\partial L}{\partial x_1} = c - \lambda = 0$

(2) $\quad \frac{\partial L}{\partial x_2} = 1 - \frac{\lambda}{x_2} = 0$

(3) $\frac{\partial L}{\partial \lambda} = b - x_1 - \ln(x_2) = 0$

¿From (1), we have $\lambda = c$. Substituting this into (2) yields:

$1 = \frac{c}{x_2}$ and so $x_2 = c$.

Substituting this result into (3) yields:

$$b - x_1 - \ln(c) = 0$$

and so:

$$x_1 = b - \ln(c)$$

The complete solution is thus: $(x_1, x_2, \lambda) = (b - \ln(c), c, c)$.

To determine whether we have obtained a minimum or maximum we can construct the bordered Hessian as follows:

$$\overline{\mathbf{H}} = \begin{bmatrix} 0 & -1 & -\frac{1}{c} \\ -1 & 0 & 0 \\ -\frac{1}{c} & 0 & \frac{1}{c} \end{bmatrix}$$

Thus, $\det(\overline{\mathbf{H}}) = 0 + (-1)^3(-1)(-\frac{1}{c}) + (-1)^4(-\frac{1}{c})0 = \frac{-1}{c} < 0$ and we have a minimum because $(-1)^m \det \overline{\mathbf{H}}_r > 0$ where $m = 1$ and $r = 2$.

Optimization with Multiple Constraints

Question 7

The Lagrangean function is:

$$L(x_1, x_2, x_3, \lambda_1, \lambda_2) = x_1 + x_2 + x_3 + \lambda_1(1 - x_1^2 - x_2^2) + \lambda_2(1 - x_2^2 - x_3^2)$$

The stationary points are found by solving the first-order necessary conditions:
(1) $\frac{\partial L}{\partial x_1} = 1 + \lambda_1(-2x_1) = 0$
(2) $\frac{\partial L}{\partial x_2} = 1 + \lambda_1(-2x_2) + \lambda_2(-2x_2) = 0$
(3) $\frac{\partial L}{\partial x_3} = 1 + \lambda_2(-2x_3) = 0$
(4) $\frac{\partial L}{\partial \lambda_1} = 1 - x_1^2 - x_2^2 = 0$
(5) $\frac{\partial L}{\partial \lambda_2} = 1 - x_2^2 - x_3^2 = 0$

Our strategy will be to first eliminate the Lagrange multipliers from (2) using (1) and (3). From (1) $\lambda_1 = \frac{1}{2x_1}$ and from (3) $\lambda_2 = \frac{1}{2x_3}$. Thus, substituting λ_1 and λ_2 into (2) we obtain:

(6) $1 = \frac{x_2}{x_1} + \frac{x_2}{x_3}$

We can divide by x_1 and x_3 because if these variables were zero, conditions (1) and (3) would not hold. The next step is to use the information in (4) and (5), i.e.,

(7) $1 - x_1^2 - x_2^2 = 1 - x_2^2 - x_3^2 \Rightarrow x_1 = \pm x_3$

It would seem that there are two possible relationships between x_1 and x_3. If, however, we substitute $x_1 = -x_3$ into (6) and simplify we obtain the result that 1=0, which is clearly false. Thus, $x_1 = x_3$. From (6), this implies

$$x_1 = 2x_2$$

Substituting this expression into (4) gives us

$$1 - 4x_2^2 - x_2^2 = 0$$
$$\Rightarrow x_2 = \pm\sqrt{\frac{1}{5}}.$$

Thus, there are only two possible solutions:

$$x^* = \left(\sqrt{\tfrac{4}{5}}, \sqrt{\tfrac{1}{5}}, \sqrt{\tfrac{4}{5}} \right)$$
$$\Rightarrow y = \sqrt{5} \; ; \text{ and}$$

$$x^{**} = \left(-\sqrt{\tfrac{4}{5}}, -\sqrt{\tfrac{1}{5}}, -\sqrt{\tfrac{4}{5}} \right)$$
$$\Rightarrow y = -\sqrt{5}.$$

We now verify whether the second-order conditions are satisfied by checking the signs of the leading bordered principal minors of the bordered Hessian. The bordered Hessian for this problem is given below:

$$\overline{\mathbf{H}} = \begin{pmatrix} 0 & 0 & -2x_1 & -2x_2 & 0 \\ 0 & 0 & 0 & -2x_2 & -2x_3 \\ -2x_1 & 0 & -2\lambda_1 & 0 & 0 \\ -2x_2 & -2x_2 & 0 & -2\lambda_1 - 2\lambda_2 & 0 \\ 0 & -2x_3 & 0 & 0 & -2\lambda_2 \end{pmatrix}$$

Substitution of x^* into (1) and (3) reveals that $\lambda_1 = \lambda_2 = \frac{1}{2\sqrt{\frac{4}{5}}}$ and substitution of x^{**} into (1) and (3) reveals that $\lambda_1 = \lambda_2 = -\frac{1}{2\sqrt{\frac{4}{5}}}$. Substituting in the values for x^* and x^{**} with the corresponding values for the Lagrangean multipliers yields the following:

solution 1 - x^*

$$\overline{\mathbf{H}} = \begin{pmatrix} 0 & 0 & -2\sqrt{\frac{4}{5}} & -2\sqrt{\frac{4}{5}} & 0 \\ 0 & 0 & 0 & -2\sqrt{\frac{1}{5}} & -2\sqrt{\frac{4}{5}} \\ -2\sqrt{\frac{4}{5}} & 0 & -\frac{1}{\sqrt{\frac{4}{5}}} & 0 & 0 \\ -2\sqrt{\frac{1}{5}} & -2\sqrt{\frac{1}{5}} & 0 & -\frac{2}{\sqrt{\frac{4}{5}}} & 0 \\ 0 & -2\sqrt{\frac{4}{5}} & 0 & 0 & -\frac{1}{\sqrt{\frac{4}{5}}} \end{pmatrix}$$

solution 2 - x^{**}

$$\overline{\mathbf{H}} = \begin{pmatrix} 0 & 0 & 2\sqrt{\frac{4}{5}} & 2\sqrt{\frac{4}{5}} & 0 \\ 0 & 0 & 0 & 2\sqrt{\frac{1}{5}} & 2\sqrt{\frac{4}{5}} \\ 2\sqrt{\frac{4}{5}} & 0 & \frac{1}{\sqrt{\frac{4}{5}}} & 0 & 0 \\ 2\sqrt{\frac{1}{5}} & 2\sqrt{\frac{1}{5}} & 0 & \frac{2}{\sqrt{\frac{4}{5}}} & 0 \\ 0 & 2\sqrt{\frac{4}{5}} & 0 & 0 & \frac{1}{\sqrt{\frac{4}{5}}} \end{pmatrix}.$$

In this case there is only one leading bordered principal minor to check as $n - m = 3 - 2 = 1$. It can be verified that at x^*, $\det(\overline{\mathbf{H}}_r) = -28.62$. Thus $(-1)^r \det(\overline{\mathbf{H}}) = (-1)^r(-28.62) > 0$ because $r = m + 1 = 3$. Thus x^* satisfies the sufficient condition for a local maximum. Similarly at x^{**} $|\overline{\mathbf{H}}_\mathbf{r}| = 28.62$ and because $(-1)^m \det(\overline{\mathbf{H}}) = (-1)^2(28.62) > 0$, x^{**} is a local minimum.

Chapter 6

Optimum Value Functions and the Envelope Theorem

Objectives

The questions in this chapter will help the reader apply duality theory to economic problems and understand the significance of the optimal value function and the Envelope Theorem. Readers who can correctly answer all the questions should be able to:

1. Apply Roy's Identity to obtain Marshallian demands (Questions 1 and 2).

2. Use the Envelope Theorem to obtain Roy's Identity (Question 3).

3. Apply Shephard's Lemma to derive input demands (Questions 4 and 5).

4. Use the Envelope Theorem to obtain Shephard's Lemma (Question 6).

5. Apply Hotelling's Lemma to derive output supply and input demands (Questions 7 and 8).

6. Use the Envelope Theorem to obtain Hotelling's Lemma (Question 9).

7. Use the Envelope Theorem to interpret the Lagrangean multiplier (Question 10).

8. Use and appreciate the importance of the Envelope Theorem (Question 11).

Review

In this chapter we examine the properties of the optimal value function. This quantity arises naturally in any optimization problem: it is the objective function evaluated at optimal values of the choice variables. Thus if $\mathbf{x}^*(\alpha)$ is the solution vector to the following problem

$$\max f(\mathbf{x}, \alpha) \text{ s.t. } g(\mathbf{x}, \alpha) = 0,$$

then the *optimal value function V* is derived by substituting the optimal values for x, defined as x^* and which are a function of the parameters, into the original objective function. The derived function $V(\alpha)$ is called the optimal value, maximum value or *indirect objective function* and is a function only of the parameters which are exogenous to the economic agent, i.e.,

$$V(\alpha) = f(x^*(\alpha)) = \max\{f(\mathbf{x}, \alpha) : g(\mathbf{x}, \alpha) = 0\}.$$

The optimal value function for a firm minimizing its costs subject to an output constraint is called the cost function while the optimal value function for a firm maximizing its profits subject to a given production technology is called the profit function. For an individual maximizing utility subject to an income constraint the optimal value function is called the indirect utility function while the optimal value function of an individual minimizing her expenditures subject to a utility constraint is called the expenditure function.

An important question concerning optimal value functions is: what happens when the parameters change? The tool for examining this question is the *Envelope Theorem*: this states that the partial derivative of the optimal value function with respect to a parameter equals the partial derivative of the associated Lagrangean function when the choice variables and Lagrangean multipliers are held equal to their optimal values, i.e.,

$$\frac{\partial V(\alpha)}{\partial \alpha} = \frac{\partial f(\mathbf{x}^*, \alpha)}{\partial \alpha} + \lambda^* \cdot \frac{\partial g(\mathbf{x}^*, \alpha)}{\partial \alpha},$$

where x^* and λ^* are the optimal values of the choice variables and the Lagrange multipliers. The Envelope theorem states that around the solution point the solution values $\mathbf{x}^*$ will be insensitive to changes in α.

The Envelope theorem can be used to derive a number of important comparitive static results and to determine the curvature of optimal value functions. In addition, from optimal value functions one can derive the choice variables as a function of the parameters. These derivations are called duality results and are particularly useful in consumer and production theory and when estimating supply and demand functions. Standard results of duality theory include *Shephard's lemma* which is used to obtain input demands from a cost function, *Hotelling's lemma* which is used to obtain output supply and input demands from a profit function and *Roy's identity* which gives the Marshallian demands from an individual's indirect utility function. These results are stated below.

Shephard's Lemma

$$\text{demand for input } i = \frac{\partial C(w, y)}{\partial w_i}$$

where $C(w, y)$ is the cost function, w is a vector of input prices and y is output.

Hotelling's Lemma

$$\text{supply of output } j = \frac{\partial \Pi(p, w)}{\partial p_j}$$

$$\text{demand for input } i = -\frac{\partial \Pi(p, w)}{\partial w_i}$$

where $\Pi(p, w)$ is the profit function, p is a vector of output prices and w is a vector of input prices.

Roy's Identity

$$\text{Marshallian demand for good } i \; = \; \frac{-\frac{\partial V(p,m)}{\partial p_i}}{\frac{\partial V(p,m)}{\partial m}}$$

where $V(p,m)$ is the indirect utility function, p is a vector of prices of the goods and m is the consumer's budget.

Further Reading

A thorough yet accessible introduction to the topic of optimal value functions is Birchenall and Grout [6], (chapters 10-12), which contains many useful examples. Another good reference is Varian [35] (chapters 1-3 and the appendix), which also provides a broad overview of both consumer theory and the theory of the firm. We also recommend Baldani et al. [2] (chapters 13 and 14) which provides a nice review of the theory with important applications. A concise and more mathematical review of optimal value functions and the Envelope Theorem is to be found in Sydsæter [31], (chapter 5 section 12). Silberberg [29] (chapter 7), Dixit (chapter 5) and Sydsæter and Hammond [32] (chapter 18.7) also provide very good descriptions of the Envelope Theorem and its proof and are highly recommended.

For an introduction to duality theory see Deaton and Muellbauer [14] (chapter 2) for consumer theory and Chambers [12] (chapters 1 to 4) for production theory. For a definitive overview see Blackorby [8] et al.

Chapter 6 - Questions

Roy's Identity

Question 1

For the utility function,

$$U = x_1 x_2^2$$

where x_1 and x_2 are the quantities of good 1 and 2, respectively,

(a) Obtain the maximum value function when the consumer faces income M where $M > 0$.

(b) Use Roy's Identity to obtain the Marshallian demands for x_1^* and x_2^*.

Question 2

An individual has the indirect utility function

$$V(p_1, p_2, M) = \frac{M^{\alpha+\beta}}{p_1^\alpha p_2^\beta}$$

where p_1 is the price of good one, p_2 is the price of good two, and M is the consumer's income. Derive the Marshallian demand function for x_1.

Question 3

For the consumer's utility maximization problem defined below derive Roy's Identity from the corresponding indirect utility function.

$$\max_{x_1, x_2} U(x_1 x_2)$$

subject to:

$$p_1 x_1 + p_2 x_2 \leq M$$

where p_1 and p_2 are the prices of goods x_1 and x_2 and M is the consumer's budget.

Shephard's Lemma

Question 4

Given an expenditure function: $E = r x_1 + w x_2$, and the constraint: $y \leq \min[x_1, x_2]$.

(a) Obtain the minimum value function for the given expenditure function.

(b) Use Shephard's Lemma to derive the Hicksian or compensated demands, x_1^* and x_2^*.

Question 5

A firm has a cost function:

$$c(w_1, w_2, y) = y[aw_1 + bw_2 + 2\alpha(w_1 w_2)^{\frac{1}{2}}]$$

where w_1 is the price of input one, w_2 is the price of input two, and y is the firm's output. Derive the firm's conditional factor demands.

Question 6

For the firm's cost minimization problem defined below derive Shephard's Lemma from the corresponding cost function.

$$\min_{x_1, x_2} \; c_1 x_1 + c_2 x_2$$

subject to:

$$f(x_1, x_2) \geq \overline{q}$$

where c_1 and c_2 are the input prices of the factors x_1 and x_2, $\overline{q}$ is the minimum output of the firm and $f(x_1, x_2)$ is the firm's production function.

Hotelling's Lemma

Question 7

A firm's profit is defined by:
$\pi = pY - wL - r\overline{K}$;
and its production function by:
$y = L^a \overline{K}^{1-a}$
where p is the output price, w is the price of labour (L) and r is the price of capital (K), fixed in supply in the short run.

(a) Obtain the maximum value function.

(b) Use Hotelling's Lemma to derive the firm's input demand, L^*.

Question 8

A firm has a profit function: $\pi(w_1, w_2, p) = \dfrac{p^{1+\alpha}}{4w_1^{\frac{\alpha}{2}} w_2^{\frac{\alpha}{2}}}$

where w_1 is the price of input one, w_2 is the price of input two, p is the output price, and $\alpha \geq 0$. Derive the firm's output supply function and input demand functions.

Question 9

For the firm's profit maximization problem defined below derive Hotelling's Lemma from the corresponding profit function.

$$\max_{x_1, x_2} \; \pi = pq - c_1 x_1 - c_2 x_2$$

subject to:

$$f(x_1, x_2) = q$$

where p is the output price, c_1 and c_2 are the input prices of the factors x_1 and x_2, q is the firm's output and $f(x_1, x_2)$ is the firm's production function.

Interpretation of the Lagrangean Multiplier

Question 10

Interpret the Lagrangean multiplier for the consumer's utility maximization problem defined below

$$\max_{x_1, x_2} U(x_1 x_2)$$

subject to:

$$p_1 x_1 + p_2 x_2 = M$$

where p_1 and p_2 are the prices of goods x_1 and x_2 and M is the consumer's budget.

Envelope Theorem

Question 11

Suppose there is a firm owned and operated by its employees who wish to maximize the profit per employee. The firm's problem, given below, is to maximize the profit per employee with respect to the amount of output (y), the amount of capital (k) and the number of employees (n).

$$\text{Max } \Pi = \frac{Py - W_k k}{n}$$

Subject to:

$$f(k, n) = y$$

where $P > 0$ is the output price, $W > 0$ is the price per unit of k, and $f(k, n)$ is the firm's production function. Assume the profit function $\Pi^*(P, W_k)$ exists and is differentiable.

What is $\frac{\partial \Pi^*(P, W_k)}{\partial P}$? Explain your answer.

Chapter 6 - Solutions

Roy's Identity

Question 1

(a) Forming the Lagrangean yields:

$$L(x_1, x_2, \lambda) = x_1 x_2^2 + \lambda(M - p_1 x_1 - p_2 x_2)$$

The necessary first-order conditions for a maximum are:

(1) $\quad \frac{\partial L}{\partial x_1} = x_2^2 - p_1 \lambda = 0$

(2) $\quad \frac{\partial L}{\partial x_2} = 2x_1 x_2 - p_2 \lambda = 0$

(3) $\quad \frac{\partial L}{\partial \lambda} = M - p_1 x_1 - p_2 x_2 = 0$

We may note that because $M > 0$ and U is monotonically increasing in x_1 and x_2 and given that if $x_1 = 0$ or $x_2 = 0$ it must be that $U = 0$, both x_1 and x_2 must both be positive at the optimum.

From (1) and (2), we can eliminate λ as follows

$$\frac{x_2^2}{p_1} = \frac{2x_1 x_2}{p_2}$$

$$\Rightarrow \quad x_2 = \frac{2p_1 x_1}{p_2}$$

Substituting this into (3), we get

$$x_1 \cdot (p_1 + 2p_1) = M$$

Thus, $x_1^* = \frac{M}{3p_1}$. Substituting into the expression for x_2 we obtain:

$$x_2^* = \frac{2p_1}{p_2} \cdot \frac{M}{3p_1} = \frac{2M}{3p_2}$$

Substituting x$_1^*$ and x_2^* into the original objective function will yield the *indirect utility function*:

$$V(p_1, p_2, M) = \frac{M}{3p_1} \cdot (\frac{2M}{3p_2})^2$$

$$V(p_1, p_2, M) = \frac{4M^3}{27p_1 p_2^2}$$

(b)
Roy's Identity states that:

$$x_i = -\frac{\frac{\partial V}{\partial p_i}}{\frac{\partial V}{\partial M}}$$

Thus, we have:

$$x_1 = -\frac{\frac{-108M^3p_2^2}{(27p_1p_2^2)^2}}{\frac{12M^2}{27p_1p_2^2}}$$

Cancelling $27p_1p_2^2$ yields:

$$x_1 = \frac{108M^3p_2^2}{12M^2 \cdot 27p_1p_2^2}$$

More cancelling yields:

$$x_1 = \frac{9M}{27p_1} = \frac{M}{3p_1}$$

We also have:

$$x_2 = -\frac{\frac{-216M^3p_1p_2}{(27p_1p_2^2)^2}}{\frac{12M^2}{27p_1p_2^2}}$$

Cancelling the $27p_1p_2^2$ term yields:

$$x_2 = \frac{216M^3p_1p_2}{12M^2 \cdot 27p_1p_2^2}$$

More cancelling yields:

$$x_2 = \frac{18M}{27p_2} = \frac{2M}{3p_2}$$

These are the same results we derived earlier from the original maximization.

Question 2

$$V(p_1, p_2, M) = \frac{M^{\alpha+\beta}}{p_1^\alpha p_2^\beta}$$

Using Roy's Identity, i.e.,

$$x_1^* = -\frac{\frac{\partial V}{\partial p_1}}{\frac{\partial V}{\partial M}}$$

where in this case

$$-\frac{\partial V}{\partial p_1} = \frac{\alpha p_1^{\alpha-1} p_2^\beta M^{\alpha+\beta}}{(p_1^\alpha p_2^\beta)^2}$$

$$\frac{\partial V}{\partial M} = \frac{(\alpha+\beta)M^{\alpha+\beta-1}}{p_1^\alpha p_2^\beta}$$

Putting the terms together:

$$x_1^* = \frac{\alpha p_1^{\alpha-1} p_2^{\beta} M^{\alpha+\beta}}{(p_1^{\alpha} p_2^{\beta})^2} \cdot \frac{p_1^{\alpha} p_2^{\beta}}{(\alpha+\beta) M^{\alpha+\beta-1}}$$

$$\Rightarrow \quad x_1^* = \frac{\alpha p_1^{\alpha-1} p_2^{\beta} M^{\alpha+\beta}}{(\alpha+\beta) M^{\alpha+\beta-1} p_1^{\alpha} p_2^{\beta}}$$

$$\Rightarrow \quad x_1^* = \frac{\alpha p_1^{\alpha-1} p_2^{\beta} M}{(\alpha+\beta) p_1^{\alpha} p_2^{\beta}}$$

$$\Rightarrow \quad x_1^* = \frac{\alpha p_1^{\alpha} p_2^{\beta} M}{(\alpha+\beta) p_1 p_1^{\alpha} p_2^{\beta}}$$

$$\Rightarrow \quad x_1^* = \frac{\alpha M}{(\alpha+\beta) p_1}$$

Question 3

The solution to the consumer's utility maximization problem are the Marshallian demand functions which may be defined as

$$x_1^* = x_1(p_1, p_2, M)$$
$$x_2^* = x_2(p_1, p_2, M)$$

Substituting x_1^* and x_2^* into the consumer's utility function we obtain the indirect utility function defined as $U(x_1^*, x_2^*) = V(p_1, p_2, M)$. The Envelope Theorem states that the partial derivative of the optimal value function with respect to a parameter equals the partial derivative of the associated Lagrangean function holding the choice variables and Lagrangean multipliers at their optimal values.

The Lagrangean function of the consumer's problem is

$$L = U(x_1, x_2) + \lambda(M - p_1 x_1 - p_2 x_2)$$

Thus, from the Envelope Theorem

$$\frac{\partial v(p_1, p_2, M)}{\partial p_1} = -\lambda^* x_1^*$$
$$\frac{\partial v(p_1, p_2, M)}{\partial p_2} = -\lambda^* x_2^*$$
$$\frac{\partial v(p_1, p_2, M)}{\partial M} = \lambda^*$$

where * indicates the optimal values of the choice variables or the Lagrangean multiplier.

Forming the negative ratio of the partial derivative of the indirect utility function with respect to price over its partial derivative with respect to income we obtain Roy's Identity, i.e.,

$$-\frac{\frac{\partial V(p_1,p_2,M)}{\partial p_1}}{\frac{\partial V(p_1,p_2,M)}{\partial M}} = \frac{\lambda^* x_1^*}{\lambda^*} = x_1^*$$

$$-\frac{\frac{\partial V(p_1,p_2,M)}{\partial p_1}}{\frac{\partial V(p_1,p_2,M)}{\partial M}} = \frac{\lambda^* x_2^*}{\lambda^*} = x_2^*$$

Shephard's Lemma

Question 4

(a) We may note that the constraint will be binding, because E is increasing in both x_1 and x_2. We may further note that the production function implies that $x_1 = x_2$ irrespective of the level of output, y, or prices r and w. This is because if $x_1 < x_2$ the firm's output would still be $y = x_1$, i.e., $y = min[x_1, x_2]$, but the firm would still incur the extra cost of paying for $x_2 - x_1$ units of x_2 at the price w per unit. Thus, the firm would only use its inputs such that $x_1 = x_2$.

$$c(r, w, y) = y \cdot (r + w)$$

(b) We can now use Shephard's Lemma, which states that the compensated demands are given by:

$$x_i^* = \frac{\partial c}{\partial w_i}$$

Thus, we have: $x_1 = y$ and $x_2 = y$, which agrees with our earlier result that $x_1 = x_2$.

Question 5

Using Shephard's Lemma

$$\frac{\partial c}{\partial w_1} = ya + y\alpha w_2 \sqrt{w_1 w_2}$$

$$\Rightarrow \frac{\partial c}{\partial w_1} = ya + y\alpha \sqrt{\frac{w_2}{w_1}}$$

By the same procedure

$$\frac{\partial c}{\partial w_2} = yb + y\alpha \sqrt{\frac{w_1}{w_2}}$$

Question 6

The solution to the firm's cost minimization problem are the input demand functions which may be defined as

$$x_1^* = x_1(c_1, c_2, \overline{q})$$
$$x_2^* = x_2(c_1, c_2, \overline{q})$$

Substituting x_1^* and x_2^* into the firm's minimization problem we obtain the cost function defined as $c_1 x_1^* + c_2 x_2^* = C(c_1, c_2, \overline{q})$. The Envelope Theorem states that the partial derivative of the optimal value function with respect to a parameter equals the partial derivative of the associated Lagrangean function holding the choice variables and Lagrangean multipliers at their optimal values.

The Lagrangean function of the firm's problem is

$$L = c_1 x_1 + c_2 x_2 + \lambda(\overline{q} - f(x_1, x_2))$$

Thus, from the Envelope Theorem we obtain Shephard's Lemma

$$\frac{\partial C(c_1, c_2, \overline{q})}{\partial c_1} = x_1^*$$
$$\frac{\partial C(c_1, c_2, \overline{q})}{\partial c_2} = x_2^*$$

where * indicates the optimal values of the choice variables.

Hotelling's Lemma

Question 7

Because the constraint holds as an equality, we can substitute for y in the firm's objective function. We can now maximize the following with respect to L.

$$\pi = pL^a \overline{K}^{1-a} - wL - r\overline{K}$$

The first order condition is as follows

$$\frac{\partial \pi}{\partial L} = apL^{a-1}\overline{K}^{1-a} - w = 0$$

Thus:

$$L^{a-1}\overline{K}^{1-a} = \frac{w}{ap}$$

$$\Rightarrow L^{a-1} = (\frac{w}{ap}) \cdot \overline{K}^{a-1}$$

$$\Rightarrow \quad L^* = (\frac{w}{ap})^{\frac{1}{a-1}} \cdot \overline{K}$$

If we substitute L^* into our objective function we will obtain the firm's maximized profit function in terms of prices and the restricted input, $\overline{K}$, i.e.,

$$\pi(r, w, p, \overline{K}) = p(\frac{w}{ap})^{\frac{a}{a-1}}\overline{K} - w^{\frac{a}{a-1}}(\frac{1}{ap})^{\frac{1}{a-1}}\overline{K} - r\overline{K}$$

where $ww^{\frac{1}{a-1}} = w^{\frac{a-1}{a-1}}w^{\frac{1}{a-1}} = w^{\frac{a}{a-1}}$.

(b) Hotelling's Lemma states that input demands are given by

$x_i = -\frac{\partial \pi(p,w)}{\partial w_i}$

Consequently

$$L^* = -\frac{\partial \pi(r, w, p, \overline{K})}{\partial w}$$

To find $\frac{\partial \pi}{\partial w}$ we use the chain rule of differentiation and obtain:

$$\frac{\partial \pi}{\partial w} = p \cdot \frac{a}{a-1} \cdot \frac{1}{ap} \cdot (\frac{w}{ap})^{\frac{a}{a-1}-1} \cdot \overline{K} - \frac{a}{a-1} \cdot w^{\frac{a}{a-1}-1} \cdot (\frac{1}{ap})^{\frac{1}{a-1}} \cdot \overline{K}$$

$$\Rightarrow \quad \frac{\partial \pi}{\partial w} = w^{\frac{1}{a-1}} \cdot \frac{1}{a-1} \cdot (\frac{1}{ap})^{\frac{1}{a-1}} \cdot \overline{K} - w^{\frac{1}{a-1}} \cdot \frac{a}{a-1} \cdot (\frac{1}{ap})^{\frac{1}{a-1}} \cdot \overline{K}$$

We may note that

$\frac{a}{a-1} - 1 = \frac{1}{a-1}$.

Thus

$$\frac{\partial \pi}{\partial w} = (\frac{1}{ap})^{\frac{1}{a-1}} \cdot \overline{K} \cdot w^{\frac{1}{a-1}} \cdot \frac{1-a}{a-1}$$

we may note that, $\frac{1-a}{a-1} = -1$.

Thus,

$$\frac{\partial \pi}{\partial w} = -(\frac{w}{ap})^{\frac{1}{a-1}} \cdot \overline{K}$$

such that

$$L^* = -\frac{\partial \pi}{\partial w} = (\frac{w}{ap})^{\frac{1}{a-1}} \cdot \overline{K}$$

and so we have confirmed our solution found from the first order condition.

Thus, the solution for L^* in (a) and (b) are identical.

Question 8

$$\pi(w_1, w_2, p) = \frac{p^{1+\alpha}}{4w_1^{\frac{\alpha}{2}} w_2^{\frac{\alpha}{2}}}$$

Using Hotelling's Lemma, output supply is

$$\frac{\partial \pi}{\partial p} = \frac{(1+\alpha)p^{\alpha}}{4w_1^{\frac{\alpha}{2}} w_2^{\frac{\alpha}{2}}}$$

and the input demands (using the quotient rule of differentiation) are

$$-\frac{\partial \pi}{\partial w_1} = \frac{2\alpha p^{1+\alpha} w_1^{\frac{\alpha}{2}-1} w_2^{\frac{\alpha}{2}}}{(4w_1^{\frac{\alpha}{2}} w_2^{\frac{\alpha}{2}})^2} = \frac{\alpha p^{1+\alpha}}{8w_1^{\frac{\alpha}{2}+1} w_2^{\frac{\alpha}{2}}}$$

$$-\frac{\partial \pi}{\partial w_2} = \frac{2\alpha p^{1+\alpha} w_1^{\frac{\alpha}{2}} w_2^{\frac{\alpha}{2}-1}}{(4w_1^{\frac{\alpha}{2}} w_2^{\frac{\alpha}{2}})^2} = \frac{\alpha p^{1+\alpha}}{8w_1^{\frac{\alpha}{2}} w_2^{\frac{\alpha}{2}+1}}$$

Question 9

The solution to the firm's profit maximization problem are the input demand functions and the output supply function which may be defined as

$$\begin{aligned} x_1^* &= x_1(p, c_1, c_2) \\ x_2^* &= x_2(p, c_1, c_2) \\ q^* &= f(x_1^*, x_2^*) \end{aligned}$$

Substituting q^*, x_1^* and x_2^* into the firm's maximization problem we obtain the profit function defined as $pq^* - c_1 x_1^* - c_2 x_2^* = \pi(p, c_1, c_2)$. The Envelope Theorem states that the partial derivative of the optimal value function with respect to a parameter equals the partial derivative of the associated Lagrangean function holding the choice variables and Lagrangean multipliers at their optimal values.

The Lagrangean function of the firm's problem is

$$L = pq - c_1 x_1 - c_2 x_2 + \lambda(q - f(x_1, x_2))$$

Thus, from the Envelope Theorem we obtain Hotelling's Lemma

$$\begin{aligned} \frac{\partial \pi(p, c_1, c_2)}{\partial p} &= q^* \\ -\frac{\partial \pi(p, c_1, c_2)}{\partial c_1} &= x_1^* \\ -\frac{\partial \pi(p, c_1, c_2)}{\partial c_2} &= x_2^* \end{aligned}$$

where * indicates the optimal values of the choice variables.

Interpretation of the Lagrangean Multiplier

Question 10

The solution to the consumer's utility maximization problem are the Marshallian demand functions which may be defined as

$$x_1^* = x_1(p_1, p_2, M)$$
$$x_2^* = x_2(p_1, p_2, M)$$

Substituting x_1^* and x_2^* into the consumer's utility function we obtain the indirect utility function defined as $U(x_1^*, x_2^*) = V(p_1, p_2, M)$. The Envelope Theorem states that the partial derivative of the optimal value function with respect to a parameter equals the partial derivative of the associated Lagrangean function holding the choice variables and Lagrangean multipliers at their optimal values.

The Lagrangean function of the consumer's problem where the variables are held at their optimal values is

$$L = U(x_1^*, x_2^*) + \lambda^*(M - p_1 x_1^* - p_2 x_2^*)$$

Thus, from the Envelope Theorem

$$\frac{\partial v(p_1, p_2, M)}{\partial M} = \lambda^*$$

where * indicates the optimal value of the Lagrangean multiplier.

Thus, the Lagrangean multiplier approximates the increase in the optimal level of utility from a marginal increase in the consumer's budget. It can be interpreted, therefore, as the marginal utility of income.

Envelope Theorem

Question 11

We may apply the Envelope Theorem that the derivative of the profit function with respect to P is the same as the derivative of the original objective function where the choice variables are held constant at their optimal values, i.e.,

$$\frac{\partial \pi^*(P, W)}{\partial P} = \frac{\partial([Py^* - W_k K^*]/n^*)}{\partial P} = \frac{y^*}{n^*}$$

Thus $\frac{\partial \pi^*(P, W)}{\partial P}$ is the firm's optimal output per employee.

Chapter 7

Linear Programming

Objectives

The questions in this chapter provide an introduction to linear programming. Readers who are able to answer all of the questions should be able to:

1. Specify simple economic problems in both the primal and dual (Questions 1,6).

2. Solve maximization problems using the simplex method (Questions 1 and 2).

3. Solve minimization problems using the simplex method (Question 6).

4. Transform problems from the primal to dual and define the notion of complementary slackness (Question 6).

5. Know when problems are infeasible (Question 3), unbounded (Question 4), or degenerate (Question 5).

Review

Problems in economics that involve an objective function and constraints that are linear in the unknown variables are linear programming problems. These types of problems may involve inequality constraints and non-negativity constraints on some or all of the variables. Every linear programming problem is characterized by an objective function (maximized or mimimized), activities to achieve the objective and a set of resources or constraints. In general, linear programming is a normative tool in that it suggests how resources should be used to maximize or minimize a given objective function. For instance, linear programming can be used to show a farmer how to maximize revenue from a choice of crops with a given set of resources or show an individual how to minimize the cost of a diet while ensuring an acceptable level of nutrition.

The approach is best illustrated using an example. Suppose a farmer wishes to maximize total revenue or the value of her crops subject to constraints such as the land, labor or machinery. If there were only two crops—potatoes and corn—valued at \$1.00 and \$2.00 a unit and only eight units of land and three units of labor available we can construct a linear programming problem provided we know how much land and labor it takes to produce one unit of each of the two crops. For instance, if we know it takes two units of land and one unit of labor to produce one unit of potatoes and five units of land and one unit of labor to produce one unit of corn we can formulate the linear programming problem. If we define potatoes as x_1 and corn as x_2, the problem is as follows:

$$\max_{x_1, x_2} \ x_1 + 2x_2$$

subject to:

$$2x_1 + 5x_2 \leq 8$$

$$x_1 + x_2 \leq 3$$

$$x_1, x_2 \geq 0$$

Whatever the farmer decides to produce she cannot use more land and labor than is available. Any mix of potatoes and corn that satisfies the contraints is called *feasible* but only the mix that satisfies the constraints **and** maximizes total revenue is *optimal*.

Linear programming problems with only three or fewer variables can be solved graphically. In the example above this would require drawing the land and labor constraints while defining the horizontal axis as units of x_1 and the vertical axis as units of x_2. The constraints would represent straight lines defined by $2x_1 + 5x_2 = 8$ and $x_1 + x_2 = 3$. The horizontal intercepts are located by setting $x_2 = 0$ while the vertical intercepts are obtained by setting $x_1 = 0$. Any combination of x_1 and x_2 on or below **both** the constraint lines is in the feasible region. The *optimal* combination is determined when the highest value total revenue line just intersects a point in the feasible region that is furthest from the origin. The total revenue lines represent different total revenues for different combinations of x_1 and x_2. All the total revenue lines have the same slope of $-\frac{1}{2}$ defined by rewriting the objective function so that x_2 is a function of a given total revenue and x_1, i.e., $x_2 = \frac{\text{Total Revenue}}{2} - \frac{x_1}{2}$. The optimal solution is shown below and is $(x_1, x_2) = (2.33, 0.67)$.

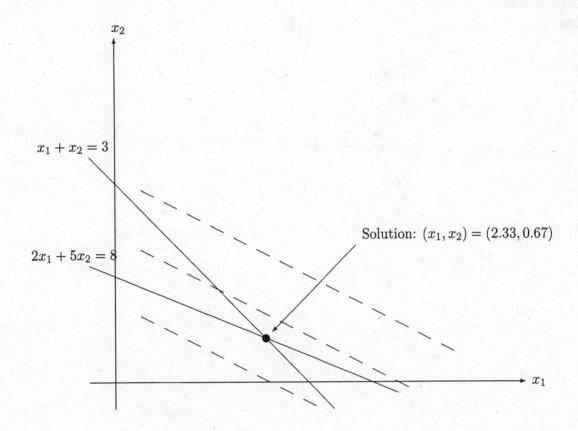

In more complex problems a number of very efficient algorithms or methods of solution can be employed. These methods, with the use of computers to help, can solve some very large and complex problems. This chapter presents the simplex method, which was developed independently by G. Danzig and L. Kantorovich, and solves a number of problems by hand. The simplex is a convex region (see chapter 3) formed by one or more linear constraints. The simplex method optimizes by moving from one feasible but not necessarily optimal solution on the simplex to another in an efficient manner until the optimal solution is found. In standard linear programming problems the first feasible solution has all the choice or *structural* variables equal to zero. The next feasible solution is found by introducing the variable which has the greatest (least) contribution to the objective function in a maximization (minimization) problem until at least one of the constraints is binding. The algorithm continues until it is not possible to improve the objective function and still ensure a feasible solution. Because a linear objective function maximized (or minimized) over a convex and closed set gives a global maximum (or minimum), we are assured that the optimal solution with the simplex method is a global maximum (or minimum).

Important concepts when using the simplex method are *slack* and *surplus* variables. Slack variables are added to constraints of the form $\leq$ such that $2x_1 + 3x_2 \leq 10$ becomes $2x_1 + 3x_2 + s_1 = 10$ where s_1 is a slack variable. Thus, if the constraint is *binding* then $s_1 = 0$. Surplus variables are substracted from $\geq$ constraints such that $2x_1 + 3x_2 \geq 10$ becomes $2x_1 + 3x_2 - s_1 = 10$. In addition, *artificial* variables are sometimes required to ensure that the *non-negativity* constraints on all variables are satisfied. For example, for the following problem:

$$\max_{x_1, x_2} \; x_1 + x_2$$

subject to:

$$2x_1 + 3x_2 \leq 10$$

$$3x_1 + x_2 \geq 6$$

$$x_1, x_2 \geq 0$$

when we add and subtract the necessary slack and surplus variables we obtain

$$\max_{x_1, x_2} \; x_1 + x_2$$

subject to:

$$2x_1 + 3x_2 + s_1 = 10$$

$$3x_1 + x_2 - s_2 = 6$$

$$x_1, x_2, s_1, s_2 \geq 0$$

If, however, the first feasible solution sets all the structural variables equal to zero then this implies from the second constraint that s_2=-6 which contradicts the non-negativity constraints and is, therefore, infeasible. To solve this difficulty we add an artificial variable such that the second constraint becomes $3x_1 + x_2 - s_2 + a_1 = 6$. To ensure that artificial variables are never included in the optimal solution they are assigned a very low (high) value in maximization (minimization) problems. Problems rewritten with slack, surplus and artificial variables are said to be in standard form.

There are many computer packages that can solve linear programming problems. When solving problems with a small number of choice variables and constraints it is possible to use a simplex tableau and apply the simplex algorithm and solve it by hand. The initial simplex tableau for the problem faced by the farmer which we solved graphically is:

i	c_b	x_b	$c_j =$	x_1	x_2	s_1	s_2	b_i	θ_i
			1 2 0 0						
1	0	s_1		2	5	1	0	8	$\frac{8}{5}$
2	0	s_2		1	1	0	1	3	3
		z_j		0	0	0	0		
		$c_j - z_j$		1	2	0	0		

The variables in the x_b column are currently in the *basis* or solution while their contribution to the objective function is in the column c_b. The c_j row gives the contribution of each of the variables to the objective function in the problem. Thus, x_1 has a c_j of one but s_1—the slack variable for the first constraint—has a c_j value of 0. The z_j row values equal the elements in the rows above multiplied by the corresponding c_b values. Thus, the z_j in column x_1 in the initial tableau is $(2 \times 0) + (1 \times 0)$ and column x_2 is $(5 \times 0) + (1 \times 0)$. The elements in the initial tableau correspond to the coefficients that multiply the variables when the problem is written in standard form and the elements in the b_i column are the total resources.

For a maximization problem, new variables enter into the basis whenever the $c_j - z_j$ values of those variables in the basis are non-zero and if variables not in the basis have a positive $c_j - z_j$ value. The pivot column for introducing a new variable into the basis is set by the rule that one picks the variable with the **highest** $c_j - z_j$ value. For minimization problems we choose the variable with the **lowest** $c_j - z_j$.

The pivot row for determining the outgoing variable is determined by the lowest positive $\theta_i = \frac{b_i}{a_{i,j}}$ where the $a_{i,j}$ term is the pivot element common to both the pivot column and row. In the tableau above, the incoming variable is x_2 because it has the highest $c_j - z_j$ value and the outgoing variable is s_1 because it has the lowest θ_i value. The pivot element is 5. The θ_i value determines how many units of the incoming variable (x_2) can be allocated by setting the outgoing variable (s_1) to zero. Once the outgoing variable has been replaced by the incoming variable in the x_b column, we apply elementary row operations to transform the tableau (including the b_i column) such that the pivot element equals one and all other elements in the pivot column equal zero. This involves the following elementary row operations:

(1) $\frac{1}{5}$row$_1$

(2) -(new)row$_1$+row$_2$

We then recalculate the z_j and $c_j - z_j$ values using the transformed elements of the tableau and continue the process until all variables in the basis have a zero value and variables not in the basis have non-positive values. The optimal value of the variables in the final simplex tableau are found in the b_i column.

Difficulties that may arise with linear programming problems include *infeasibility*, *degeneracy*, and *unboundedness*. Infeasibility arises when it is impossible to satisfy the constraints and arises from a misspecification of the problem. Degeneracy arises from *cycling* in the algorithm from one feasible solution to another without reaching the optimum. Certain rules can be applied to the simplex method to avoid this problem. Unboundedness arises when the constraints are insufficient to constrain the solution to finite values of the structural variables.

An important concept in linear programming is *duality*. Duality enables one to solve a *primal* problem such as maximizing revenue by solving an equivalent *dual* problem that minimizes the resource cost. For the general maximization problem defined as follows:

$$\max_x c_1 x_1 + c_2 x_2 \cdots + c_n x_n$$

subject to:

$$a_{11}x_1 \quad +a_{12}x_2 \cdots a_{1n}x_n \leq \quad b_1$$
$$\vdots \qquad \vdots \qquad \vdots$$
$$a_{m1}x_1 \quad +a_{m2}x_2 \cdots a_{mn}x_n \leq \quad b_m$$

$$x_1, x_2, \cdots x_n \geq 0$$

The dual problem can be written as follows:

$$\min_{\lambda} b_1\lambda_1 + b_2\lambda_2 \cdots + b_m\lambda_m$$

subject to:

$$a_{11}\lambda_1 \quad +a_{21}\lambda_2 \cdots a_{m1}\lambda_m \geq \quad c_1$$

$$a_{n1}\lambda_1 \quad +a_{n2}\lambda_2 \cdots a_{nm}\lambda_m \geq \quad c_n$$

$$\lambda_1, \lambda_2, \cdots \lambda_m \geq 0$$

The transformation involves changing from maximization (minimization) of n x variables to minimization (maximization) of m λ dual variables, reversing the inequalities, transposing the matrix of coefficients on the left hand side of the constraints, and switching the coefficients of the structural variables with the resources on the right hand side of the constraints. It follows that if the primal problem is infeasible but the dual is feasible then the dual will be unbounded.

The value of the objective function at the optimum for the primal and dual problems are identical. If the ith constraint in the primal problem holds as a strict inequality at the optimum then the ith dual variable is zero. Further, if a variable x_i is positive valued at the optimal solution in the primal problem, the jth dual constraint holds as a strict equality. These results are known as *complementary slackness*. If the primal problem is infeasible but the dual is feasible then the dual will be unbounded.

Further Reading

Readers interested in obtaining an introduction to linear programming are encouraged to read Black and Bradley [7] (chapter 12), Chiang [13] (chapters 19 and 20), Lambert [24] (chapter 5 section 5.5), Haeussler and Paul [21] (chapter 7), Rowcroft [26] (chapter 13) and Sydsæter and Hammond [32] (chapter 19). Sydsæter and Hammond [32] (chapter 19.5) also offer a complete but easy to follow explanation of complementary slackness.

For a more complete treatment there are many excellent texts on linear programming including Bradley et al. [9], Dorfman et al. [16] and Wu and Coppins [38]. Both Bradley et al. [9] and Wu and Coppins [38] are highly recommended while Dorfman et al. [16] provides a number of examples of how linear programming may be applied in economics. Schrage [28] provides a text and software for the readers interested in using personal computers to solve linear and quadratic programming problems.

Chapter 7 - Questions

Maximization

Question 1

•A firm produces two types of ice cream, using cream and food additives. To produce variety A, the firm requires 3 units of cream and 3 units of food additives. To produce variety B, the firm requires 4 units of cream but only 1 unit of food additives. At most, the firm has 12 units of cream and 6 units of food additives, and both varieties sell for $1.00 per unit.

(a) Formulate the primal problem in canonical form for the above firm, assuming that it wishes to maximize its total revenue.

(b) Formulate the dual problem for (a).

(c) Solve the problem using the simplex method, and illustrate your answer with a diagram.

Question 2

Solve the following problem using the simplex method and draw the feasible region on a graph:

$$\max_{x_1, x_2} f = 3x_1 + 2x_2$$

subject to:

$$5x_1 + 4x_2 \leq 20$$

$$2x_1 + 4x_2 \leq 16$$

$$x_1, x_2 \geq 0$$

Infeasibility

Question 3

Solve the following linear programming problem.

$$\min_{y_1, y_2} f = -3y_1 + 4y_2$$

subject to:

$$y_1 + y_2 \geq 2$$

$$-y_1 - 3y_2 \geq 3$$

$$y_1, y_2 \geq 0$$

Unboundedness

Question 4

Solve the following problem.

$$\max_{x_1,x_2} f = 3x_1 + 4x_2$$

subject to:

$$x_1 + x_2 \geq 2$$

$$x_1 - 3x_2 \leq 3$$

$$x_1, x_2 \geq 0$$

Degeneracy

Question 5

$$\max_{x_1,x_2} \Pi = 3x_1 + x_2$$

subject to:

$$x_1 + 2x_2 \leq 4$$

$$2x_1 + x_2 \leq 6$$

$$x_1 + x_2 \leq 3$$

$$x_1, x_2, s_1, s_2, s_3 \geq 0$$

Minimization

Question 6

A farmer produces cauliflower and cabbages by using land, labor, and capital. To produce one unit of cauliflower he needs one unit of land, three units of labor, and one unit of capital. To produce one unit of cabbages he needs one unit of land, one unit of labor, and two units of capital. The farmer has at her disposal 100 units of land, 200 units of labor, and 180 units of capital. The price the farmer receives per unit of cauliflower and cabbage is, respectively, $5.00 and $1.00 unit.

(a) Formulate the linear programming problem of the farmer assuming she wishes to maximize her total revenue.

(b) Formulate the dual problem to the primal problem given in part (a).

(c) Solve the dual problem and compare with the solution to the primal problem.

Chapter 7 - Solutions

Maximization

Question 1

(a) Primal Problem

$$\max_{x_1, x_2} TR = x_1 + x_2$$

subject to:

$$3x_1 + 4x_2 \leq 12$$

$$3x_1 + x_2 \leq 6$$

$$x_1, x_2 \geq 0$$

(b) Dual Problem
Applying the transformation to a dual problem we obtain

$$\min_{\lambda_1, \lambda_2} V = 12\lambda_1 + 6\lambda_2$$

subject to:

$$3\lambda_1 + 3\lambda_2 \geq 1$$

$$4\lambda_1 + \lambda_2 \geq 1$$

$$\lambda_1, \lambda_2 \geq 0$$

(c) Solution
To solve the problem in the primal form, we first write the problem in *standard form* whereby we add a slack variable to each $\leq$ constraint:

$$\max_{x_1, x_2} TR = x_1 + x_2$$

subject to:

$$3x_1 + 4x_2 + s_1 = 12$$

$$3x_1 + x_2 + s_2 = 6$$

$$x_1, x_2, s_1, s_2 \geq 0$$

We can now write out the initial simplex tableau:

	$c_j =$		1	1	0	0		
i	c_b	x_b	x_1	x_2	s_1	s_2	b_i	θ_i
1	0	s_1	3	4	1	0	12	4
2	0	s_2	3	1	0	1	6	2
		z_j	0	0	0	0		
		$c_j - z_j$	1	1	0	0		

The variables in the x_b column are currently in the *basis* or solution. The simplex algorithm enters new variables into the basis whenever the $c_j - z_j$ values of those variables in the basis are non-zero and if variables not in the basis have a positive $c_j - z_j$ value. The pivot column for introducing a new variable into the basis is set by the rule that one picks the variable with the highest $c_j - z_j$ value. The pivot row for determining the outgoing variable is determined by the lowest positive $\theta_i = \frac{b_i}{a_{i,j}}$ where the $a_{i,j}$ term is the pivot element common to both the pivot column and row. This determines how many units of the incoming variable (x_j) can be allocated by setting the outgoing variable (x_i) to zero.

Following the rules of the algorithm, the incoming variable may be either x_1 or x_2 since both have the highest $c_j - z_j$ value. We choose x_1. The outgoing variable is the slack variable s_2 with the lowest θ_i value equal to 2. The second tableau is derived by making the pivot element equal to one removing the incoming variable, x_1, from the remaining rows. The elementary row operations are as follows:

(1) $\frac{1}{3}$row$_2$

(2) -3(new)row$_2$ + row$_1$

	$c_j =$		1	1	0	0		
i	c_b	x_b	x_1	x_2	s_1	s_2	b_i	θ_i
1	0	s_1	0	3	1	-1	6	2
2	1	x_1	1	$\frac{1}{3}$	0	$\frac{1}{3}$	2	6
		z_j	1	$\frac{1}{3}$	0	$\frac{1}{3}$		
		$c_j - z_j$	0	$\frac{2}{3}$	0	$-\frac{1}{3}$		

Because x_2 has a positive $c_j - z_j$ value the procedure must be repeated. The new elementary row operations are as follows:

(1) $\frac{1}{3}$ row$_1$

(2) $-\frac{1}{3}$ (new)row$_1$ + row$_2$

	$c_j =$		1	1	0	0		
i	c_b	x_b	x_1	x_2	s_1	s_2	b_i	θ_i
1	1	x_2	0	1	$\frac{1}{3}$	$-\frac{1}{3}$	2	
2	1	x_1	1	0	$-\frac{1}{9}$	$\frac{4}{9}$	$\frac{4}{3}$	
		z_j	1	1	$\frac{2}{9}$	$\frac{1}{9}$		
		$c_j - z_j$	0	0	$-\frac{2}{9}$	$-\frac{1}{9}$		

The algorithm is complete as all variables in the basis have zero $c_j - z_j$ values and the variables not in the basis have non-positive $c_j - z_j$ values.

The solution is obtained from the b_i column where $(x_1, x_2) = (\frac{4}{3}, 2)$. Total revenue is found by multiplying the c_j values by the b_i in the final simplex tableau to obtain $3\frac{1}{3}$. The shadow price of the input cream, or the increase in total revenue by increasing the quantity of cream available from 12 to 13 units is found in the z_j row and equals $\frac{2}{9}$. Similarly, the shadow price for food additives is found to be $\frac{1}{9}$.

A graphical solution is provided in below. The simplex is a convex region formed by one or more linear constraints where the ridge of the simplex is a line on the boundary and a vertex is where two or more ridges intersect. The simplex method involves an efficient search among the possible vertices to find the vertex that maximizes the objective function. For the problem in question 1, this vertex $(x_1, x_2) = (\frac{4}{3}, 2)$ is reached when the highest value total revenue line just intersects the feasible region at a point furthest from the origin.

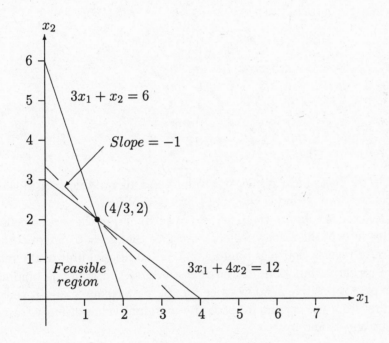

Question 2

We begin by writing the problem in the *standard form* whereby we add a slack variable to each $\leq$ constraint:

$$\max_{x_1, x_2} f = 3x_1 + 2x_2$$

subject to:

$$5x_1 + 4x_2 + s_1 = 20$$

$$2x_1 + 4x_2 + s_2 = 16$$

$$x_1, x_2, s_1, s_2 \geq 0$$

The initial simplex tableau becomes,

i	c_b	x_b	$c_j =$	3 x_1	2 x_2	0 s_1	0 s_2	b_i	θ_i
1	0	s_1		5	4	1	0	20	4
2	0	s_2		2	4	0	1	16	8
		z_j		0	0	0	0		
		$c_j - z_j$		3	2	0	0		

Choose as an incoming variable, the pivot column, the variable with the highest $c_j - z_j$ value. Choose as the outgoing variable, pivot row, the variable with the lowest θ_i value. Use elementary row operations to remove x_1 from row$_2$ and set the pivot element equal to one.

(1) $\frac{1}{5}$ row$_1$

(2) $-2 \cdot$ (the new)row$_1$ + row$_2$

i	c_b	x_b	$c_j =$	3 x_1	2 x_2	0 s_1	0 s_2	b_i	θ_i
1	3	x_1		1	$\frac{4}{5}$	$\frac{1}{5}$	0	4	
2	0	s_2		0	$\frac{12}{5}$	$-\frac{2}{5}$	1	8	
		z_j		3	$\frac{12}{5}$	$\frac{3}{5}$	0		
		$c_j - z_j$		0	$-\frac{2}{5}$	$-\frac{3}{5}$	0		

Because the variables now in the basis have zero $c_j - z_j$ values and all variables not in the basis have non-positive $c_j - z_j$ values we have obtained the final simplex tableau.

The solution may be read from the b_i column where $(x_1, s_2) = (4, 8)$. From the second constraint if $x_1 = 4$ and there are 8 surplus units of the resource then it must be that $x_2 = 0$. Substituting the value of x_1 into the first constraint reveals that it is binding, i.e., an additional unit of the resource would change the value of the objective function. Given that the first constraint is binding and the second is not, we would expect a positive shadow price for the first resource and a zero price for the second resource. Reading from the z_j row we note the marginal value of an extra unit of resource one and two is, respectively, $\frac{3}{5}$ and 0.

A graph of the solution is provided:

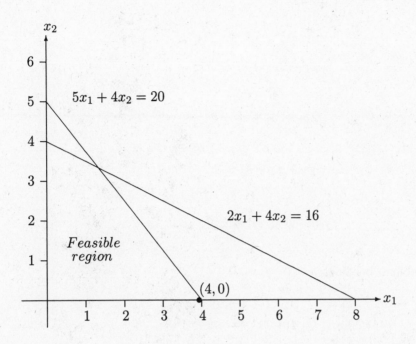

Infeasibility

Question 3

This problem can be shown to be *infeasible*. If we multiply the second constraint by -1, we obtain:

$$y_1 + 3y_2 \leq -3$$

which cannot be satisfied if $y_1 \geq 0$ and $y_2 \geq 0$. In more complicated problems, infeasibility can be found if there are one or more artificial variables in the basis which are positive. It should be noted that if the primal problem is infeasible but the dual is feasible, then the dual is unbounded.

Unboundedness

Question 4

We may observe from this problem that if we set $x_1 = 0$ we can increase the objective function without bound by increasing x_2 while still satisfying the constraints. The problem, therefore, is *unbounded*.

Degeneracy

Question 5

We first write the problem in *standard form* whereby we add a slack variable to each $\leq$ constraint:

$$\max_{x_1, x_2} \Pi = 3x_1 + x_2$$

subject to:

$$x_1 + 2x_2 + s_1 = 4$$

$$2x_1 + x_2 + s_2 = 6$$

$$x_1 + x_2 + s_3 = 3$$

$$x_1, x_2, s_1, s_2, s_3 \geq 0$$

The initial simplex tableau becomes:

		$c_j =$	3	1	0	0	0		
i	c_b	x_b	x_1	x_2	s_1	s_2	s_3	b_i	θ_i
1	0	s_1	1	2	1	0	0	4	4
2	0	s_2	2	1	0	1	0	6	3
3	0	s_3	1	1	0	0	1	3	3
		z_j	0	0	0	0	0		
		$c_j - z_j$	3	1	0	0	0		

The pivot column or incoming variable is determined by the highest $c_j - z_j$ value (x_1). The rule for choosing the pivot row or outgoing variable is to pick the row with the lowest positive θ_i value. In this problem both the s_2 and s_3 rows have identical θ_i values. Whenever this arises the problem is said to be *degenerate*. The problem occurs whenever the b_i in the m technical constraints are expressible as a linear combination of fewer than m columns. In this problem it means that the vector formed by the b_i column of dimension (3×1) is a linear combination of fewer than 3 of the column vectors formed by x_1, x_2, s_1, s_2, s_3 in the initial simplex tableau. This does not mean there is not a feasible solution to the problem but that at the solution at least three 3 constraints including non-negativity constraints will be binding simultaneously.

One procedure to solve a degenerate problem is to apply the following rules. First, pick as the incoming variable or pivot column the variable furthest to the left in the simplex tableau that has a positive $c_j - z_j$ value. Second, pick the outgoing variable with the lowest positive θ_i and if there is a tie among the outgoing variables, choose the variable located furthest to the left in the simplex tableau. The second rule should still be applied when the $\theta_i = 0$. Applying this rule we choose as the incoming variable x_1 and the outgoing variable s_2. Performing the elementary row operations such that x_1 is removed for rows 1 and 3 and that the pivot element equals 1.

(1) $\frac{1}{2}$ row$_2$

(2) -(new)row$_2$+ row$_1$

(3) -(new)row$_2$+ row$_3$

i	c_b	x_b	x_1	x_2	s_1	s_2	s_3	b_i	θ_i
		$c_j =$	3	1	0	0	0		
1	0	s_1	0	$\frac{3}{2}$	1	$-\frac{1}{2}$	0	1	4
2	0	x_1	1	$\frac{1}{2}$	0	$\frac{1}{2}$	0	3	3
3	0	s_3	0	$\frac{1}{2}$	0	$-\frac{1}{2}$	1	0	3
		z_j	3	1	0	1	0		
		$c_j - z_j$	0	0	0	-1	0		

The problem is solved as all variables in the basis have a zero $c_j - z_j$ value and all variables not in the basis do not have a positive $c_j - z_j$ value. The solution can be read from the b_i column as $(s_1, x_1, s_3) = (1, 3, 0)$. This indicates the first constraint is not binding but the third constraint is binding. Substituting $x_1 = 3$ into the constraints reveals that the second constraint holds as a strict equality and that $x_2 = 0$ such that the non-negativity constraint on x_2 is binding.

Minimization

Question 6

(a) **Primal Problem**

$$\max_{x_1, x_2} 5x_1 + x_2$$

subject to:

$$x_1 + x_2 \leq 100$$

$$3x_1 + x_2 \leq 200$$

$$x_1 + 2x_2 \leq 180$$

$$x_1, x_2 \geq 0$$

(b) **Dual Problem**

$$\min_{\lambda_1, \lambda_2, \lambda_3} 100\lambda_1 + 200\lambda_2 + 180\lambda_3$$

subject to:

$$\lambda_1 + 3\lambda_2 + \lambda_3 \geq 5$$

$$\lambda_1 + \lambda_2 + 2\lambda_3 \geq 1$$

$$\lambda_1, \lambda_2, \lambda_3 \geq 0$$

(c) Solution

To obtain the initial simplex tableau to the dual problem rewrite the problem in *standard form* by subtracting surplus variables. In addition, if the non-negativity constraints are to be satisfied an artificial variable must be added to each $\geq$ constraint. In minimization (maximization) problems the artificial variables are assigned an arbitrarily large positive (negative) value, defined by M or $-M$, to ensure that the variables are not included in the final simplex tableau.

$$\min_{\lambda_1, \lambda_2, \lambda_3} 100\lambda_1 + 200\lambda_2 + 180\lambda_3$$

subject to:

$$\lambda_1 + 3\lambda_2 + \lambda_3 - s_1 + a_1 = 5$$

$$\lambda_1 + \lambda_2 + 2\lambda_3 - s_2 + a_2 = 1$$

$$\lambda_1, \lambda_2, \lambda_3, s_1, s_2, a_1, a_2 \geq 0$$

For the initial simplex tableau we choose the artificial variables to be in the basis:

i	c_b	x_b	$c_j =$ 100 λ_1	200 λ_2	180 λ_3	0 s_1	0 s_2	M a_1	M a_2	b_i	θ_i
1	M	a_1	1	3	1	-1	0	1	0	5	$\frac{5}{3}$
2	M	a_2	1	1	2	0	-1	0	1	1	1
		z_j	$2M$	$4M$	$3M$	$-M$	$-M$	M	M		
		$c_j - z_j$	$100 - 2M$	$200 - 4M$	$180 - 3M$	M	M	0	0		

Because this is a minimization problem, we choose the incoming variable with the **lowest** $c_j - z_j$ value (λ_2). The pivot row, as with a maximization problem, is chosen by the variable with the lowest positive θ_i value (a_2). Using elementary row operations we must eliminate λ_2 from row 1. For convenience, we will drop the artificial variable a_2 from the second simplex tableau as it no longer appears in the basis.

(1) $-3\text{row}_2 + \text{row}_1$

i	c_b	x_b	$c_j =$ 100 λ_1	200 λ_2	180 λ_3	0 s_1	0 s_2	M a_1	b_i	θ_i
1	M	a_1	-2	0	-5	-1	3	1	2	$\frac{2}{3}$
2	200	λ_2	1	1	2	0	-1	0	1	-1
		z_j	$-2M + 200$	200	$-5M + 400$	$-M$	$3M - 200$	M		
		$c_j - z_j$	$2M - 100$	0	$5M - 220$	M	$200 - 3M$	0		

The final simplex tableau is obtained by performing the following elementary row operations observing s_2 is the incoming variable and a_1 is the outgoing variable. Because the artificial variable a_1 no longer appears in the basis we may drop it from the tableau.

(1) $\frac{1}{3}$ row$_1$

(2) (new)row$_1$+ row$_2$

		$c_j =$	100	200	180	0	0		
i	c_b	x_b	λ_1	λ_2	λ_3	s_1	s_2	b_i	θ_i
1	0	s_2	$-\frac{2}{3}$	0	$-\frac{5}{3}$	$-\frac{1}{3}$	1	$\frac{2}{3}$	
2	200	λ_2	$\frac{1}{3}$	1	$\frac{1}{3}$	$-\frac{1}{3}$	0	$\frac{5}{3}$	
		z_j	$\frac{200}{3}$	200	$\frac{200}{3}$	$-\frac{200}{3}$	0		
		$c_j - z_j$	$\frac{100}{3}$	0	$\frac{340}{3}$	$\frac{200}{3}$	0		

Because all variables in the basis have zero $c_j - z_j$ values and all variables not in the basis have positive $c_j - z_j$ values we have obtained the final simplex tableau. The solution is $(\lambda_1, \lambda_2, \lambda_3) = (0, \frac{5}{3}, 0)$ and $(s_1, s_2) = (0, \frac{2}{3})$ with the objective function $= \frac{1000}{3}$.

Observation of the primal problem in (a) reveals that the solution is to produce as much x_1 given the constraints. The only binding constraint is the second constraint such that the maximum value of x_1 is $\frac{200}{3}$ if $x_2 = 0$. This gives the same value of the objective function as the solution to the dual problem. This is a general result; provided that optimal and feasible solutions exist for the primal and dual, the value of the objective functions will be identical. We may further note that because the first and third constraints are non-binding then the shadow price of land and capital is zero. In the case of the labor constraint, an additional unit of the input can increase the output of x_1 by $\frac{1}{3}$ unit which in turn can increase the objective function by $\frac{5}{3}$. We may observe that the value of λ_i in the solution to the dual problem is their shadow prices. Consequently, $\lambda_2 = \frac{5}{3}$ and $\lambda_1 = \lambda_3 = 0$. The solution illustrates the *complementary slackness* conditions in that if the ith constraint in the primal problem holds as a strict inequality then the ith dual variable is zero.

Chapter 8

Nonlinear Programming

Objectives

The questions in this chapter will enable the reader to solve some nonlinear programming problems and to understand when the conditions for a solution are necessary and when they are both necessary and sufficient. Readers who are able to answer all the questions should be able to:

1. Solve general maximization problems using the Kuhn-Tucker conditions (Questions 1, 3 and 4).

2. Transform a minimization problem to a maximization problem (Question 2).

3. Apply an algorithm to solve quadratic programming problems involving a quadratic objective function and linear constraints (Question 5).

4. Use the Kuhn-Tucker conditions to solve linear programming problems and understand the symmetry between the Lagrangean multipliers and the dual variables (Question 6).

5. Apply and understand the Arrow-Enthoven conditions or quasiconcave programming criteria and know when the Kuhn-Tucker conditions are both necessary and sufficient (Question 7).

6. Know the difference between the necessary and sufficient conditions in nonlinear programming (Question 8).

Review

Many problems in economics cannot be expressed simply in terms of a linear objective function and constraints. Problems where the unknown variables are nonlinear in the objective function and/or constraints may be solved through nonlinear programming. Unlike linear programming which uses a very efficient algorithm to obtain the optimum, nonlinear programming often requires a case by case approach to obtain the optimum.

One approach to solving nonlinear programming problems is to use the Kuhn-Tucker conditions. For the maximization problem with m constraints and n variables with $x = x_1, x_2 \cdots x_n$ is given below:

$$\max_x f(x)$$

subject to:

$$g^i(x) \leq c_i \text{ for all } i = 1, 2, \cdots m$$
$$x_j \geq 0 \text{ for all } j = 1, 2 \cdots n$$

The corresponding Lagrangean may be written as follows:

$$L = f(x) + \sum_{i=1}^{m} \lambda_i(c_i - g^i(x))$$

The Kuhn-Tucker conditions for a maximum are:

(1) $\frac{\partial L}{\partial x_j} = f_j(x) - \sum_{i=1}^{m} \lambda_i g_j^i(x) \leq 0; \ x_j \cdot \left(\frac{\partial L}{\partial x_j}\right) = 0$ for all j

(2) $\frac{\partial L}{\partial \lambda_i} = c_i - g^i(x) \geq 0; \ \lambda_i \cdot \left(\frac{\partial L}{\partial \lambda_i}\right) = 0$ for all i

(3) $x_j \geq 0$ for all j

(4) $\lambda_i \geq 0$ for all i

where $f_j(x)$ and $g_j^i(x)$ are the partial derivative of the objective function ($f(x)$) and constraint function ($g^i(x)$) with respect to the variable x_j. Minimization problems can be transformed into maximization problems by multiplying the objective function and the technological constraints ($g^i(x) \leq c_i$ constraints) by -1 and reversing the inequalities in the technological constraints.

The method of solution using the Kuhn-Tucker conditions consists of eliminating alternatives until a solution is found. Under certain conditions, the Kuhn-Tucker conditions are both *necessary* and *sufficient* for a maximum thus ensuring that a solution that satisfies the conditions is indeed the maximum. These conditions known as *quasiconcave programming* or the *Arrow-Enthoven conditions* are defined below for the general maximization problem.

$$\max_x \ Z = f(x_1,, x_2, \cdots, x_n)$$

subject to:

$$g^i(x_1, x_2, \cdots, x_n) \leq c_i \text{ for all } i = 1, \cdots, m$$
$$x_j \geq 0 \text{ for all } j = 1, \cdots, n$$

The quasiconcave programming or Arrow-Enthoven conditions are satisfied if for all $x \geq 0$

1. $f(x_1, \cdots, x_n)$ is differentiable and quasiconcave.

2. Each $g^i(x_1, \cdots, x_n)$ is differentiable and quasiconvex.

3. The *constraint qualification* is satisfied. Provided that the objective function and constraint functions are differentiable and defined over a convex set the constraint qualification is satisfied if **either** one of the following conditions is true:

   (a) If all the constraint functions are convex and there exists a vector $\underline{x}$ such that $g^i(x_1, x_2, \cdots, x_n) < c_i$ for all i for $x_1, x_2, \cdots, x_n \geq 0$.

   (b) The matrix of first-order partial derivatives of all the **binding** or active constraints (constraints where $g^i(x_1, x_2, \cdots, x_n) = c_i$ at the maximum) when evaluated at the maximum x^* has maximum rank.

   The constraint qualification is automatically satisfied if **all** the constraints are linear.

4. If any **one** of the following is satisfied at a vector $\underline{x}$ which satisfies the Kuhn-Tucker maximum conditions given that $f(x_1, \cdots, x_n)$ is differentiable and quasiconcave and $g^i(x_1, \cdots, x_n)$ is differentiable and quasiconvex.

   (a) $\frac{\partial f(\underline{x})}{\partial x_j} < 0$ for at least one variable x_j.

   (b) $\frac{\partial f(\underline{x})}{\partial x_j} > 0$ for at least one variable x_j without violating the constraints.

   (c) All the second-order partial derivatives of $f(\underline{x})$ exist although they may be all zero.

   (d) The function $f(x_1, \cdots, x_n)$ is concave.

The last condition is a *sufficient* condition such that if a vector $\underline{x}$ satisfies the condition then it will be a global maximum. However, a vector $\underline{x}$ may not satisfy the sufficient condition and still be a global maximum. If a vector $\underline{x}$ satisfies all four of the Arrow-Enthoven conditions we are assured that the solution is a unique global maximum. If the Arrow-Enthoven conditions are not satisfied but the constraint qualification is satisfied then the Kuhn-Tucker are only *necessary* conditions for an optimum provided that the objective function and constraint functions are continuously differentiable.

If the constraint qualification condition is satisfied, $f()$ is assumed to be differentiable and concave, and if each $g^i()$ is differentiable and convex then the conditions are called the criteria for *concave programming*. It is a stronger set of conditions and also ensures that the Kuhn-Tucker conditions are both necessary and sufficient. The usefulness of the Arrow-Enthoven conditions is the following: if the maximization problem satisfies the Arrow-Enthoven conditions then the problem has a unique solution which will uniquely satisfy the the Kuhn-Tucker conditions.

Certain types of nonlinear programming are more easily solved using algorithms. For example, *quadratic programming* problems can be solved using a modified form of the simplex method (see chapter 7) provided that the objective function involves a quadratic function and is strictly concave (see chapter 3) and all the constraints are linear.

To apply the algorithm the problem must first be written in the following form,

$$\max_{x} V = \sum_{i=1}^{n} c_j x_j - \frac{1}{2} \sum_{j=1}^{n} \sum_{k=1}^{n} q_{jk} x_j x_k$$

subject to:

$$\sum_{j=1}^{n} a_{ij} x_j \leq b_i \text{ for } i = 1, \cdots, m$$

$$x_j \geq 0 \text{ for } j = 1, \cdots, n$$

where m equals the number of technical constraints, n equals the number of variables, x_j are the choice variables, c_j are the coefficients of the choice variables in the objective function, b_i are the resource constraints and q_{ij} are coefficients used to transform the original objective function into the necessary form.

The method of solution involves transforming the maximization problem into a minimization problem and solving it using a modified version of the simplex algorithm (see chapter 7). Using the previously defined variables, the general minimization problem is as follows:

$$\min_{a_j} Z = \sum_{i=1}^{n} a_j$$

subject to:

$$\sum_{k=1}^{n} q_{jk}x_k + \sum_{i=1}^{m} t_{ij}y_{n+i} - y_j + a_j = c_j \text{ for } j = 1, \cdots, n$$

$$\sum_{i=1}^{n} a_{ij}x_j + s_{n+i} = b_i \text{ for } i = 1, \cdots, m$$

$$x_j y_j = 0 \text{ for } j = 1, \cdots, n$$

$$s_{n+j}y_{n+j} = 0 \text{ for } j = n+1, \cdots, n+m$$

$$x_j, y_j \geq 0 \text{ for } j = 1, \cdots, n$$

$$s_{n+j}, y_{n+j} \geq 0 \text{ for } j = n+1, \cdots, n+m$$

$$a_j \geq 0 \text{ for } j = 1, \cdots, n$$

where a_j are artificial variables needed to generate the initial feasible solution, $s_{n+1,\cdots n+m}$ are slack variables for each of the technical constraints, and $y_{1,\cdots,n}$ are surplus variables used to ensure the Kuhn-Tucker conditions are satisfied.

Further Reading

There are a number of references that the reader may find useful on nonlinear programming including: Beavis and Dobbs [4], Chiang [13] (chapter 21), Glaister [19] (pp. 184-188), Lambert [24] (chapter 5), Sydsæter [31] (chapter 5 sections 13-16), Sydsæter and Hammond [32] (chapter 18.8-18.10) Rowcroft [26] (chapter 18), Wismer and Chattergy [36], and Wu and Coppins [38] (chapter 12). Beavis and Dobbs [4] (chapter 2, section 2.3) provide an excellent summary of the necessary and sufficient conditions and the constraint qualification while Sydsæter and Hammond [32] (chapter 18.8) provide a useful introduction to the topic. Chiang [13], Baldani et al. [2] (chapter 12) and Lambert [24] provide a number of additional problems and exercises that may be useful to the interested reader while Rowcroft [26] provides a nice introduction to the topic. Wismer and Chattergy [36] present a number of solved problems in nonlinear programming including quadratic programming while Wu and Coppins [38] provide a good introduction to quadratic programming.

Chapter 8 - Questions

Kuhn-Tucker Conditions and Maximization

Question 1

Solve the following maximization problem to obtain the Marshallian demand functions, x_1^* and x_2^*. In the problem, the consumer's income is defined by M and P_1 and P_2 are, respectively, the price of the consumption goods x_1 and x_2.

$$\max_{x_1,x_2} U = x_1^a x_2^b$$

subject to:

$$p_1 x_1 + p_2 x_2 \leq M$$
$$x_1 \leq c x_2$$
$$x_1, x_2 \geq 0$$

where M, p_1, p_2, a, b, and c are all positive numbers.

Question 2

Minimize the following function

$$\min_{x_1,x_2} V = 2x_1 + 2x_2$$

subject to:

$$x_1^2 - 4x_1 + x_2 \geq 0$$
$$-2x_1 - 3x_2 \geq -10$$
$$x_1, x_2 \geq 0$$

Question 3

Solve the following maximization problem to obtain the Marshallian demand functions, x_1^* and x_2^*. In the problem, the consumer's income is defined by M and P_1 and P_2 are, respectively, the price of the consumption goods x_1 and x_2.

$$\max_{x_1,x_2} U = x_1 + x_2$$

subject to:

$$P_1 x_1 + P_2 x_2 \leq M$$
$$x_1, x_2 \geq 0$$

where M, P_1, and P_2 are all positive numbers.

Question 4

Maximize the following function

$$\max_{x_1, x_2} Y = x_1^{\frac{1}{3}} + x_2^{\frac{1}{3}}$$

subject to:

$$x_1 \leq 4$$
$$x_2 \leq 6$$
$$x_1 + x_2 \leq 5$$
$$x_1, x_2 \geq 0$$

Quadratic Programming

Question 5

A monopolist wishes to maximize total revenue. It produces two outputs, (x_1, x_2) and faces the following demands for its products, $x_1 = 20 - 2p_1$ and $x_2 = 20 - 4p_2$ where p_1 and p_2 are, respectively, the prices of the two goods. To produce one unit of x_1 the monopolist must use one unit of land and one unit of capital and to produce one unit of x_2 requires two units of land and one unit of capital. The firm has available 10 units of land and 5 units of capital.

(a) Specify the firm's short-run maximization problem.

(b) Determine the output of x_1 and x_2, the total revenue of the firm, and the amount of land and capital used at the optimum.

Kuhn-Tucker Conditions and Dual Variables

Question 6

A firm wishes to maximize its short-run net profits. It is able to produce three goods which, respectively, contribute 20, 20, and 8 dollars per unit to net profits. Each unit of the first good requires 10 units of labor and 6 units of capital, each unit of the second good requires 2 units of labor and 4 units of capital, and the third good requires 3 units of labor and one unit of capital. In the short run, the firm is constrained to only 20 units of labor and 14 units of capital.

(a) Solve the firm's problem using the Kuhn-Tucker conditions noting the optimal quantities of the three goods that should be produced, the quantity of labor and capital used, and the value of the Lagrangean multipliers.

(b) Obtain the dual of the firm's linear programming short-run profit maximization problem. Solve the dual via the simplex method or by graphical means. Compare the solution of the dual variables to the value of the Lagrangean multipliers.

(c) In the general case where the objective function and constraints are linear, compare the Lagrangean multipliers obtained from the Kuhn-Tucker conditions to the shadow prices in the solution to the dual linear programming problem.

Quasiconcave Programming

Question 7

$$\max_{x_1, x_2} \ (x_1 - 1)^2 + (x_2 - 2)^2$$

subject to:

$$x_1 + x_2 \leq 4$$
$$x_2 \leq 2$$
$$x_1, x_2 \geq 0$$

(a) Determine whether the Arrow-Enthoven conditions are satisfied for the above maximization problem.

(b) Verify that the solution $(x_1, x_2) = (4,0)$ satisfies the Kuhn-Tucker conditions.

(c) Determine whether the Kuhn-Tucker conditions are satisfied at the $(x_1, x_2) = (1,2)$. Comment on the result.

Necessary and Sufficient Conditions for a Maximum

Question 8

For a general maximization problem with one variable and a linear constraint show that the *minimum* value of the constrained objective function can satisfy the Kuhn-Tucker conditions.

Chapter 8 - Solutions

Kuhn-Tucker Conditions and Maximization

Question 1

Forming the Lagrangean yields:

$$L = x_1^a x_2^b + \lambda_1(M - p_1 x_1 - p_2 x_2) + \lambda_2(c x_2 - x_1)$$

The Kuhn-Tucker conditions are:

(1) $\quad \frac{\partial L}{\partial x_1} = a x_1^{a-1} x_2^b - \lambda_1 p_1 - \lambda_2 \leq 0; \ x_1 \cdot \left(\frac{\partial L}{\partial x_1}\right) = 0$

(2) $\quad \frac{\partial L}{\partial x_2} = b x_1^a x_2^{b-1} - \lambda_1 p_2 + \lambda_2 c \leq 0; \ x_2 \cdot \left(\frac{\partial L}{\partial x_2}\right) = 0$

(3) $\quad \frac{\partial L}{\partial \lambda_1} = M - p_1 x_1 - p_2 x_2 \geq 0; \ \lambda_1 \cdot \left(\frac{\partial L}{\partial \lambda_1}\right) = 0$

(4) $\quad \frac{\partial L}{\partial \lambda_2} = -x_1 + c x_2 \geq 0; \ \lambda_2 \cdot \left(\frac{\partial L}{\partial \lambda_2}\right) = 0$

(5) $\quad x_1 \geq 0$

(6) $\quad x_2 \geq 0$

(7) $\quad \lambda_1 \geq 0$

(8) $\quad \lambda_2 \geq 0$

Note that if either $x_1 = 0$ or $x_2 = 0$, $U = 0$. We can thus safely rule out that (5) and (6) are binding constraints (ie. we can assume that $x_1, x_2 > 0$). We may also assume that the budget constraint is binding given $c > 0$. This is because if not all the consumer's income is spent it will always be possible to increase x_1 and/or x_2 and, thereby, increase the objective function.

Under the above reasoning, therefore, there are two possible alternatives:

(A) $\quad \lambda_1, \lambda_2 > 0$

(B) $\quad \lambda_1 > 0, \lambda_2 = 0$

We will consider each alternative in turn.

<u>Alternative A</u> $(\lambda_1, \lambda_2 > 0)$

By assumption $x_1, x_2 > 0$, thus (1) to (4) hold as equalities. If (4) is binding, then $x_1 = c x_2$. Substituting this into (3) yields:

$$x_1 = \frac{cM}{cp_1 + p_2}$$

$$x_2 = \frac{M}{cp_1 + p_2}$$

We can now check to see if this answer satisfies (1) and (2):

Multiply (1) by x_1 to obtain:

$$aU = (\lambda_1 p_1 + \lambda_2)x_1$$

Multiply (2) by x_2 to obtain:

$$bU = (\lambda_1 p_2 - \lambda_2 c) x_2$$

Rearranging the expressions we obtain

$$(\lambda_1 p_1 + \lambda_2) x_1 b = (\lambda_1 p_2 - \lambda_2 c) x_2 a$$

Given that $cx_2 = x_1$, we have:

$$(\lambda_1 p_1 + \lambda_2) c x_2 b = (\lambda_1 p_2 - \lambda_2 c) x_2 a$$

Rearranging yields:

$$x_2(\lambda_1 p_1 cb - \lambda_1 p_2 a) = x_2[(-\lambda_2 c)(a + b)] < 0$$

The above is true because $x_2, \lambda_1, \lambda_2, a, b, c > 0$.

Thus, $\lambda_1 p_1 cb - \lambda_1 p_2 a < 0$ which implies $\frac{a}{b} > \frac{cp_1}{p_2}$. The solution will only hold true under this condition.

Alternative B $(\lambda_1 > 0, \lambda_2 = 0)$

From (1) and (2), we have:

$$x_1 \lambda_1 p_1 b = x_2 \lambda_1 p_2 a$$

Thus: $x_1 p_1 = x_2 p_2 \frac{a}{b}$

Substituting into (3) and rearranging we eventually obtain:

$$x_1 = \frac{aM}{p_1(a + b)}$$

$$x_2 = \frac{bM}{p_2(a + b)}$$

Now, substitute these values for x_1 and x_2 into (4) to see if the constraint holds as an inequality:

$$x_1 \le c x_2$$

$$\Rightarrow \frac{aM}{p_1(a + b)} \le \frac{cbM}{p_2(a + b)}$$

$$\Rightarrow \frac{a}{p_1} \le \frac{cb}{p_2}$$

$$\Rightarrow \frac{a}{b} \leq \frac{cp_1}{p_2}$$

This second solution only holds true under the above condition.

Given that the objective function is differentiable and quasiconcave (because $a, b > 0$) and all the constraints are linear we are assured that the Kuhn-Tucker conditions are both necessary and sufficient. Thus, the complete answer is:

If $\frac{a}{b} \leq \frac{cp_1}{p_2}$ then:

$$x_1 = \frac{aM}{p_1(a+b)}$$

$$x_2 = \frac{bM}{p_2(a+b)}$$

If $\frac{a}{b} > \frac{cp_1}{p_2}$ then:

$$x_1 = \frac{cM}{cp_1 + p_2}$$

$$x_2 = \frac{M}{cp_1 + p_2}$$

Question 2

We can rewrite this problem as a maximization problem to obtain the Kuhn-Tucker conditions with which we may be more familiar by multiplying everything by -1 and reversing the inequalities.

$$\max_{x_1, x_2} Z = -2x_1 - 2x_2$$

subject to:

$$-x_1^2 + 4x_1 - x_2 \leq 0$$

$$2x_1 + 3x_2 \leq 10$$

$x_1 \geq 0$ and $x_2 \geq 0$.

Forming the Lagrangean yields:

$$L = -2x_1 - 2x_2 + \lambda_1(x_1^2 + x_2 - 4x_1) + \lambda_2(10 - 2x_1 - 3x_2)$$

The Kuhn-Tucker conditions are:

(1) $\frac{\partial L}{\partial x_1} = -2 + \lambda_1(2x_1 - 4) - 2\lambda_2 \leq 0;\ x_1 \cdot \left(\frac{\partial L}{\partial x_1}\right) = 0$

(2) $\frac{\partial L}{\partial x_2} = -2 + \lambda_1 - 3\lambda_2 \leq 0;\ x_2 \cdot \left(\frac{\partial L}{\partial x_2}\right) = 0$

(3) $\frac{\partial L}{\partial \lambda_1} = x_1^2 + x_2 - 4x_1 \geq 0;\ \lambda_1 \cdot \left(\frac{\partial L}{\partial \lambda_1}\right) = 0$

(4) $\frac{\partial L}{\partial \lambda_2} = 10 - 2x_1 - 3x_2 \geq 0;\ \lambda_2 \cdot \left(\frac{\partial L}{\partial \lambda_2}\right) = 0$

(5) $x_1 \geq 0$

(6) $x_2 \geq 0$

(7) $\lambda_1 \geq 0$

(8) $\lambda_2 \geq 0$

We note that under the condition $x_1 \geq 0, x_2 \geq 0$, the highest value that the objective function, Z, can attain in the maximization problem is zero such that $(x_1, x_2) = (0,0)$. It can easily be verified that this solution satifies the two constraints. The problem illustrates that the solution to nonlinear programming problems is *not* simply an application of a solution algorithm. The procedure often involves examining each possible candidate for a solution in turn and rejecting those alternatives that fail to satisfy the Kuhn-Tucker conditions. In the case where the Arrow-Enthoven conditions are satisfied (see question 7) then the problem will have a unique solution that will uniquely satisfy the Kuhn-Tucker conditions.

Question 3

Forming the Lagrangean yields:

$$L(x_1, x_2, \lambda) = x_1 + x_2 + \lambda(M - p_1 x_1 - p_2 x_2)$$

The Kuhn-Tucker conditions are:

(1) $\frac{\partial L}{\partial x_1} = 1 - p_1\lambda \leq 0;\ x_1 \cdot \left(\frac{\partial L}{\partial x_1}\right) = 0$

(2) $\frac{\partial L}{\partial x_2} = 1 - p_2\lambda \leq 0;\ x_2 \cdot \left(\frac{\partial L}{\partial x_2}\right) = 0$

(3) $\frac{\partial L}{\partial \lambda} = M - p_1 x_1 - p_2 x_2 \geq 0;\ \lambda \cdot \left(\frac{\partial L}{\partial \lambda}\right) = 0$

(4) $x_1 \geq 0$

(5) $x_2 \geq 0$

(6) $\lambda \geq 0$

It must be the case that $\lambda > 0$ otherwise it is possible to increase either x_1 and/or x_2 by some small amount, satisfy the constraints, and increase the objective function. Given $\lambda > 0$, there are three possible alternatives for a solution:

(A) $x_1 > 0, x_2 = 0$

(B) $x_1 > 0, x_2 > 0$

(C) $x_1 = 0, x_2 > 0$

We will consider each alternative in turn.

<u>Alternative A</u> $(x_1 > 0, x_2 = 0)$

Given the assumption that $x_2 = 0$ and that $\lambda > 0$ then from (3),

$$M - p_1 x_1 = 0$$

Thus, $x_1^* = \frac{M}{p_1}$ remembering that by assumption $x_2 = 0$.

To find the conditions in which this solution holds true, we note that from (1)

$$1 = p_1 \lambda$$

Substituting into (2):

$$p_1 \lambda \leq p_2 \lambda$$

$$\Rightarrow p_1 \leq p_2$$

<u>Alternative B</u> $(x_1 > 0, x_2 > 0)$

By assumption, we have $x_1, x_2 > 0$ and given that $\lambda > 0$, we have from (3)

$$p_1 x_1 + p_2 x_2 = M$$

From (1) and (2) we also have:

$$1 = p_1 \lambda$$

$$1 = p_2 \lambda$$

which leads to the following

$$p_1 \lambda = p_2 \lambda$$

$$\Rightarrow \quad p_1 = p_2$$

Substituting this condition into (3) we obtain:

$$p(x_1 + x_2) = M$$

$$\Rightarrow x_1 + x_2 = \frac{M}{p}$$

<u>Alternative C</u> $(x_1 = 0, x_2 > 0)$

By assumption $x_1 = 0$ and given that $\lambda > 0$ we note from (3)

$$x_2^* = \frac{M}{p_2}$$

To find the condition under which this solution holds true, we note that given $x_2 > 0$, from (2)

$$1 = p_2\lambda$$

Substituting this into (1) yields:

$$p_2\lambda \leq p_1\lambda$$

$$\Rightarrow \quad p_2 \leq p_1$$

Observing that the objective function is concave and all the constraints are linear then the Kuhn-Tucker conditions will be both necessary and sufficient. Thus the results from all three alternatives gives us the following Marshallian demand functions:

Marshallian Demand for x_1

$x_1 = \frac{M}{p_1}$, given $p_1 < p_2$.
$x_1 =$ any value from 0 to $\frac{M}{p}$, given $p_1 = p_2$.
$x_1 = 0$, given $p_1 > p_2$.

Marshallian Demand for x_2

$x_2 = \frac{M}{p_2}$, given $p_1 > p_2$.
$x_2 =$ any value from 0 to $\frac{M}{p}$, given $p_1 = p_2$.
$x_2 = 0$, given $p_1 < p_2$.

Question 4

Forming the Lagrangean yields:

$$L(x_1, x_2, \lambda_1, \lambda_2, \lambda_3) = x_1^{\frac{1}{3}} + x_2^{\frac{1}{3}} + \lambda_1(4 - x_1) + \lambda_2(6 - x_2) + \lambda_3(5 - x_1 - x_2)$$

The Kuhn-Tucker conditions are:

(1) $\quad \frac{\partial L}{\partial x_1} = \frac{1}{3}x_1^{\frac{-2}{3}} - \lambda_1 - \lambda_3 \leq 0; \ x_1 \cdot (\frac{\partial L}{\partial x_1}) = 0$

(2) $\quad \frac{\partial L}{\partial x_2} = \frac{1}{3}x_2^{\frac{-2}{3}} - \lambda_2 - \lambda_3 \leq 0; \ x_2 \cdot (\frac{\partial L}{\partial x_2}) = 0$

(3) $\quad \frac{\partial L}{\partial \lambda_1} = 4 - x_1 \geq 0; \ \lambda_1 \cdot (\frac{\partial L}{\partial \lambda_1}) = 0$

(4) $\quad \frac{\partial L}{\partial \lambda_2} = 6 - x_2 \geq 0; \ \lambda_2 \cdot (\frac{\partial L}{\partial \lambda_2}) = 0$

(5) $\quad \frac{\partial L}{\partial \lambda_3} = 5 - x_1 - x_2 \geq 0; \ \lambda_3 \cdot (\frac{\partial L}{\partial \lambda_3}) = 0$

(6) $\quad x_1 \geq 0$

(7) $\quad x_2 \geq 0$

(8) $\quad \lambda_1 \geq 0$

(9) $\quad \lambda_2 \geq 0$

(10) $\quad \lambda_3 \geq 0$

It must be the case that at least **one** of the constraints is binding; otherwise it is possible to increase either x_1 and/or x_2 by some small amount, satisfy the constraints, and increase the objective function. Before examining the alternatives for the Lagrangean multipliers (λ_1, λ_2 and λ_3) we will first examine the three possible alternatives for a solution with respect to the original choice variables:

(A) $x_1 > 0, x_2 = 0$

(B) $x_1 = 0, x_2 > 0$

(C) $x_1 > 0, x_2 > 0$

We will consider each alternative in turn.

<u>Alternative A</u> $(x_1 > 0, x_2 = 0)$

Given the assumption that $x_2 = 0$ then we would like to have the highest possible value for x_1 while still satisfying the constraints. Observation of the three constraints reveals that the highest possible value of x_1 is 4. At this value only the first constraint is binding and holds as a strict equality while the second and third constraints are strict inequalities. It follows, therefore, that only a change in the first constraint will change the value of the objective function. Thus $\lambda_1 > 0$ while $\lambda_2 = \lambda_3 = 0$.

To verify whether this is a constrained maximum we need to verify that the Kuhn-Tucker conditions are satisfied. Multiplying (2) by $x_2^{\frac{2}{3}}$ does not change the inequality and we obtain

$$\frac{1}{3} \leq x_2^{\frac{2}{3}}(\lambda_2 + \lambda_3)$$

However, from above $\lambda_2 = \lambda_3 = 0$ and by assumption $x_2 = 0$ thus (2) implies that $\frac{1}{3} < 0$ which is false. Thus we have a contradiction and this cannot be the solution because the Kuhn-Tucker conditions are both necessary and sufficient as the objective function is concave and all the contraints are linear.

<u>Alternative B</u> $(x_1 = 0, x_2 > 0)$

Given the assumption that $x_1 = 0$ then we would like to have the possible value for x_2 while still satisfying the constraints. Observation of the three constraints reveals that the highest possible value of x_2 is 5. At this value only the third constraint is binding and holds as a strict equality while the first and second constraints are strict inequalities. It follows, therefore, that only a change in the third constraint will change the value of the objective function. Thus $\lambda_3 > 0$ while $\lambda_1 = \lambda_2 = 0$.

To verify whether this is a constrained maximum we need to verify that the Kuhn-Tucker conditions are satisfied. Multiplying (1) by $x_1^{\frac{2}{3}}$ does not change the inequality and we obtain

$$\frac{1}{3} \leq x_1^{\frac{2}{3}}(\lambda_1 + \lambda_3)$$

However, from above $\lambda_1 = 0$ and by assumption $x_1 = 0$, thus (1) implies that $\frac{1}{3} < 0$ which is false. Thus we have a contradiction and this cannot be the solution because the Kuhn-Tucker conditions are both necessary and sufficient as the objective function is concave and all the contraints are linear.

<u>Alternative C</u> $(x_1 > 0, x_2 > 0)$

Under this alternative it must be the case that the third constraint is binding; otherwise it would be possible to increase either x_1 and/or x_2 by a very small amount, satisfy the constraints and increase the objective function. Thus at the constrained maximum $\lambda_3 > 0$ which implies

$$5 - x_1 - x_2 = 0$$

To determine what would be the optimal quatities of x_1 and x_2 we need to appreciate that if $x_1 > x_2$ then the marginal contribution to the objective function of a very small change in x_1 is **less than** the marginal contribution of a very small change in x_2, i.e.,

$$\frac{\partial y}{\partial x_1} < \frac{\partial y}{\partial x_2}$$

Similarly, if $x_1 < x_2$ then

$$\frac{\partial y}{\partial x_1} > \frac{\partial y}{\partial x_2}$$

Thus, if possible we will maximize the objective function when $x_1 = x_2$ provided that the third constraint is binding. This implies that the constrained maximum is $(x_1, x_2) = (\frac{5}{2}, \frac{5}{2})$. This solution provides the following values for the Lagrangean multipliers $(\lambda_1, \lambda_2, \lambda_3) = (0, 0, \frac{\frac{1}{2}}{3(5^{\frac{2}{3}})})$. Verification of the Kuhn-Tucker conditions reveals that this solution does indeed satisfy all the conditions which are both necessary and sufficient for a maximum.

Quadratic Programming
Question 5

(a)

$$\max_{x_1, x_2} \text{TR} = 10x_1 + 5x_2 - \frac{x_1^2}{2} - \frac{x_2^2}{4}$$

subject to:

$$x_1 + 2x_2 \leq 10$$
$$x_1 + x_2 \leq 6$$
$$x_1, x_2 \geq 0$$

(b)

It may be observed that this problem involves a quadratic objective function with linear constraints. If the objective function is strictly concave, there is an algorithm that permits us to solve the problem in a straight-forward manner using the simplex method in linear programming. This approach is particularly useful in more complex problems.

To use this approach we must first verify that the objective function is strictly concave. Forming the Hessian matrix of second-order partial derivatives,

$$H = \begin{bmatrix} \frac{\partial^2 y}{\partial x_1^2} & \frac{\partial^2 y}{\partial x_1 \cdot \partial x_2} \\ \frac{\partial^2 y}{\partial x_2 \cdot \partial x_1} & \frac{\partial^2 y}{\partial x_2^2} \end{bmatrix} = \begin{bmatrix} -1 & 0 \\ 0 & -\frac{1}{2} \end{bmatrix}$$

where

$$|H_1| = -1 < 0, \quad |H| = \frac{1}{2} > 0$$

Thus, the Hessian has the appropriate alternating signs of the leading principal minors and, therefore, is negative definite. This implies the function is strictly concave.

To apply the algorithm the problem must first be written in the following form,

$$\max_x V = \sum_{i=1}^{n} c_j x_j - \frac{1}{2} \sum_{j=1}^{n} \sum_{k=1}^{n} q_{jk} x_j x_k$$

subject to:

$$\sum_{j=1}^{n} a_{ij} x_j \le b_i \text{ for } i = 1, \cdots, m$$

$$x_j \ge 0 \text{ for } j = 1, \cdots, n$$

where m equals the number of technical constraints and n equals the number of variables. The maximization problem in (a) can be written in this form as follows:

$$\max_{x_1, x_2} V = 10x_1 + 5x_2 - \frac{1}{2}(x_1^2 + \frac{x_2^2}{2})$$

subject to:

$$x_1 + 2x_2 \le 10$$

$$x_1 + x_2 \le 6$$

$$x_1, x_2 \ge 0$$

where $n = 2$ and $m = 2$ and in the objective function $c_1 = 10$, $c_2 = 5$, $q_{11} = 1$, $q_{12} = q_{21} = 0$, and $q_{22} = \frac{1}{2}$. In the constraints $b_1 = 10$, $b_2 = 6$, $a_{11} = 1$, $a_{12} = 2$, $a_{21} = 1$, and $a_{22} = 1$.

The procedure is to transform the maximization problem into a minimization problem which can be solved using a modified version of the simplex algorithm. Using the previously defined variables, the general minimization problem is as follows:

$$\min_{a_j} Z = \sum_{i=1}^{n} a_j$$

subject to:

$$\sum_{k=1}^{n} q_{jk} x_k + \sum_{i=1}^{m} t_{ij} y_{n+i} - y_j + a_j = c_j \text{ for } j = 1, \cdots, n$$

$$\sum_{i=1}^{n} a_{ij} x_j + s_{n+i} = b_i \text{ for } i = 1, \cdots, m$$

$$x_j y_j = 0 \text{ for } j = 1, \cdots, n$$

$$s_{n+j} y_{n+j} = 0 \text{ for } j = n+1, \cdots, n+m$$

$$x_j, y_j \ge 0 \text{ for } j = 1, \cdots, n$$

$$s_{n+j}, y_{n+j} \ge 0 \text{ for } j = n+1, \cdots, n+m$$

$$a_j \ge 0 \text{ for } j = 1, \cdots, n$$

where a_j are artificial variables needed to generate an initial basis, $s_{n+1,\cdots n+m}$ are slack variables for each of the technical constraints, and $y_{1,\cdots,n}$ are surplus variables used to ensure the Kuhn-Tucker conditions are satisfied.

Writing the minimization problem for the problem at hand

$$\min_{a_1,a_2} Z = a_1 + a_2$$

subject to:

$$
\begin{aligned}
x_1 + t_{11}y_3 + t_{21}y_4 - y_1 + a_1 &= 10 \\
\tfrac{1}{2}x_2 + t_{12}y_3 + t_{22}y_4 - y_2 + a_2 &= 5 \\
x_1 + 2x_2 + s_3 &= 10 \\
x_1 + x_2 + s_4 &= 6 \\
x_1 y_1, x_2 y_2 &= 0 \\
s_3 y_3, s_4 y_4 &= 0 \\
x_1, x_2, y_1, y_2, y_3, y_4 &\geq 0 \\
a_1, a_2, s_3, s_4 &\geq 0
\end{aligned}
$$

The inital simplex tableau for the minimization problem can be easily derived noting that the artificial variables (a_1, a_2) and slack variables (s_3, s_4) are included in the initial *basis* or solution. The method of solution will be familiar to readers who know the simplex method in linear programming (see chapter 7).

i	c_b	x_b	x_1	x_2	y_1	y_2	y_3	y_4	s_3	s_4	a_1	a_2	b_i	θ_i
		$c_j =$	0	0	0	0	0	0	0	0	1	1		
1	1	a_1	1	0	-1	0	1	1	0	0	1	0	10	10
2	1	a_2	0	$\tfrac{1}{2}$	0	-1	1	1	0	0	0	1	5	$-$
3	0	s_3	1	2	0	0	0	0	1	1	0	0	10	10
4	0	s_4	1	1	0	0	0	0	0	0	0	0	6	6
		z_j	1	$\tfrac{1}{2}$	-1	-1	2	2	0	0	1	1		
		$c_j - z_j$	-1	$-\tfrac{1}{2}$	1	1	-2	-2	0	0	0	0		

Because this is a minimization problem, we choose the incoming variable with the **lowest** $c_j - z_j$ value. Given the constraint that $s_j y_j = 0$, however, we cannot let either y_3 or y_4 enter the basis. We, therefore, choose the next best variable x_1. The pivot row is chosen by the variable with the lowest positive θ_i value (s_4). Using elementary row operations we must eliminate x_1 from all other rows. This involves the following elementary row operations:

(1) -row$_4$ + row$_1$

(2) -row$_4$ + row$_3$

		$c_j =$	0	0	0	0	0	0	0	0	1	1		
i	c_b	x_b	x_1	x_2	y_1	y_2	y_3	y_4	s_3	s_4	a_1	a_2	b_i	θ_i
1	1	a_1	0	-1	-1	0	1	1	0	-1	1	0	4	4
2	1	a_2	0	$\frac{1}{2}$	0	-1	1	1	0	0	0	1	5	5
3	0	s_3	0	1	0	0	0	0	1	-1	0	0	4	$-$
4	0	x_1	1	1	0	0	0	0	0	1	0	0	6	$-$
		z_j	0	$-\frac{1}{2}$	-1	-1	2	2	0	-1	1	1		
		$c_j - z_j$	0	$\frac{1}{2}$	1	1	-2	-2	0	1	0	0		

The new pivot column now becomes y_4 because it has the lowest $c_j - z_j$ value and s_4 is no longer in the basis. The pivot row or outgoing variable becomes a_1 because it has the lowest positive θ_i value. The elementary row operations to eliminate y_4 from all other rows are defined below.

(1) -row$_1$ + row$_2$

		$c_j =$	0	0	0	0	0	0	0	0	1	1		
i	c_b	x_b	x_1	x_2	y_1	y_2	y_3	y_4	s_3	s_4	a_1	a_2	b_i	θ_i
1	0	y_4	0	-1	-1	0	1	1	0	-1	1	0	4	$-$
2	1	a_2	0	$\frac{3}{2}$	0	-1	0	0	0	1	-1	1	1	$\frac{2}{3}$
3	0	s_3	0	1	0	0	0	0	1	-1	0	0	4	4
4	0	x_1	1	1	0	0	0	0	0	1	0	0	6	6
		z_j	0	$\frac{3}{2}$	1	-1	0	0	0	1	-1	1		
		$c_j - z_j$	0	$-\frac{3}{2}$	-1	1	0	0	0	-1	2	0		

The pivot column now becomes x_2 because it has the lowest $c_j - z_j$ value and the pivot row is a_2 with the lowest positive θ_i value. Because the artificial variables a_1 and a_2 are no longer in the basis we drop them from the next tableau. The elementary row operations to eliminate x_2 from the other rows are given below.

(1) $\frac{2}{3}$row$_2$

(2) row$_2$ + row$_1$

(3) -(new)row$_2$ + row$_3$

(4) -(new)row$_2$ + row$_4$

		$c_j =$	0	0	0	0	0	0	0	0		
i	c_b	x_b	x_1	x_2	y_1	y_2	y_3	y_4	s_3	s_4	b_i	θ_i
1	0	y_4	0	0	$-\frac{1}{3}$	$\frac{2}{3}$	1	1	0	$-\frac{1}{3}$	$4\frac{2}{3}$	$-$
2	0	x_2	0	1	$\frac{2}{3}$	$-\frac{2}{3}$	0	0	0	$\frac{2}{3}$	$\frac{2}{3}$	$\frac{2}{3}$
3	0	s_3	0	0	$-\frac{2}{3}$	$\frac{2}{3}$	0	0	1	$-\frac{5}{3}$	$3\frac{1}{3}$	4
4	0	x_1	1	0	$-\frac{2}{3}$	$\frac{2}{3}$	0	0	0	$\frac{1}{3}$	$5\frac{1}{3}$	6
		z_j	0	0	0	0	0	0	0	0		
		$c_j - z_j$	0	0	0	0	0	0	0	0		

The problem is solved as all variables in the basis have a zero $c_j - z_j$ value while the variables not in the basis have a non-negative $c_j - z_j$ value.

The solution can be read from the b_i column such that $(x_1, x_2) = (5\frac{1}{3}, \frac{2}{3})$. The slack variable s_3 has a value of $3\frac{1}{3}$ indicating that the first constraint is non-binding and that there are $10 - 3\frac{1}{3} = 6\frac{2}{3}$ units of land used. Substitution of the optimal values x_1 and x_2 into the capital constraint reveals that 6 units of capital are used by the monopolist. Substitution of the optimal values of x_1 and x_2 into the objective function gives a maximum short-run total revenue of $42\frac{1}{3}$.

The problem could also have been solved through the use of the Kuhn-Tucker conditions by eliminating alternatives that provide a contradiction. In more complex problems, where there are many constraints, this procedure can be rather laborious.

Kuhn-Tucker Conditions and Dual Variables

Question 6

(a)

$$\max_{x_1, x_2, x_3} \Pi = 20x_1 + 20x_2 + 8x_3$$

subject to:

$$10x_1 + 2x_2 + 3x_3 \leq 20$$
$$6x_1 + 4x_2 + x_3 \leq 14$$
$$x_1, x_2, x_3 \geq 0$$

Forming the Lagrangean yields:

$$L = 20x_1 + 20x_2 + 8x_3 + \lambda_1(20 - 10x_1 - 2x_2 - 3x_3) + \lambda_2(14 - 6x_1 - 4x_2 - x_3)$$

The Kuhn-Tucker conditions are:

(1) $\frac{\partial L}{\partial x_1} = 20 - 10\lambda_1 - 6\lambda_2 \leq 0$; $x_1 \cdot (\frac{\partial L}{\partial x_1}) = 0$

(2) $\frac{\partial L}{\partial x_2} = 20 - 2\lambda_1 - 4\lambda_2 \leq 0$; $x_2 \cdot (\frac{\partial L}{\partial x_2}) = 0$

(3) $\frac{\partial L}{\partial x_3} = 8 - 3\lambda_1 - \lambda_2 \leq 0$; $x_3 \cdot (\frac{\partial L}{\partial x_3}) = 0$

(4) $\frac{\partial L}{\partial \lambda_1} = 20 - 10x_1 - 2x_2 - 3x_3 \geq 0$; $\lambda_1 \cdot (\frac{\partial L}{\partial \lambda_1}) = 0$

(5) $\frac{\partial L}{\partial \lambda_2} = 14 - 6x_1 - 4x_2 - x_3 \geq 0$; $\lambda_2 \cdot (\frac{\partial L}{\partial \lambda_2}) = 0$

(6) $x_1 \geq 0$

(7) $x_2 \geq 0$

(8) $x_3 \geq 0$

(9) $\lambda_1 \geq 0$

(10) $\lambda_2 \geq 0$

We may observe that x_1 and x_2 contribute the same in net profit per unit to the firm's objective function. However, x_2 uses less labor and capital than x_1. In the case where either of these constraints are binding then x_2 will always be preferred over x_1. We may, therefore, assume that the optimal solution will have $x_1 = 0$. In terms of the alternatives of x_2 and x_3 we have:

1. x_2 and $x_3 > 0$.

2. $x_2 > 0$ and $x_3 = 0$

3. $x_2 = 0$ and $x_3 > 0$.

Alternative 1

Given that the objective function is strictly increasing in all variables it must be the case that at least one of the constraints is binding. In the case of the first alternative, therefore, the Lagrangean multipliers may have the following values,

1. $\lambda_1 = 0$ and $\lambda_2 > 0$.

2. $\lambda_1 > 0$ and $\lambda_2 = 0$.

3. $\lambda_1 > 0$ and $\lambda_2 > 0$.

Consider the case where $\underline{\lambda_1 = 0, \lambda_2 > 0}$.
By assumption $x_1 = 0$ and $x_2, x_3 > 0$. This implies the following:

$$\text{From (2)} \quad 20 \ = \ 4\lambda_2 \Rightarrow \lambda_2 = 5$$
$$\text{From (3)} \quad 8 \ = \ \lambda_2$$

This is a contradiction so this alternative cannot be the optimum.

Consider the case where $\underline{\lambda_1 > 0, \lambda_2 = 0}$.

By assumption $x_1 = 0$ and $x_2, x_3 > 0$. This implies the following:

$$\text{From (2)} \quad 20 \ = \ 2\lambda_1 \Rightarrow \lambda_1 = 10$$
$$\text{From (3)} \quad 8 \ = \ 3\lambda_2 \Rightarrow \lambda_1 = 2\frac{2}{3}$$

This is a contradiction so this alternative cannot be the optimum.

Consider the case where $\underline{\lambda_1, \lambda_2 > 0}$.

By assumption $x_1 = 0$ and $x_2, x_3 > 0$. This implies the following:

$$\text{From (2)} \quad \lambda_1 \ = \ 10 - 2\lambda_2$$
$$\text{From (3)} \quad 8 \ = \ 3\lambda_1 + \lambda_2$$

Substituting λ_1 into the second expression above and solving we obtain $\lambda_2 = 4.4$. Substituting back into the expression for λ_1 we obtain the following values for the Lagrangean multipliers, $(\lambda_1, \lambda_2) = (1.2, 4.4)$.

To find the solution for x_2 and x_3 we note that because λ_1 and λ_2 are both greater than zero the labor and capital constraints must both be binding. The Kuhn-Tucker conditions imply the following:

$$\text{From (4)} \quad 2x_2 + 3x_3 \ = \ 20$$
$$\text{From (5)} \quad 4x_2 + x_3 \ = \ 14$$

Solving the two equations with two unknowns we obtain $(x_2, x_3) = (2.2, 5.2)$. Substituting the values for the Lagrangean multipliers and the optimum quantities we can verify that the Kuhn-Tucker conditions are satisfied. Substituting the optimal quantities of the goods into the objective function we obtain a net profit of 85.6.

Alternative 2

This alternative assumes $x_1, x_3 = 0$ and $x_2 > 0$. Observation of the labor and capital constraints reveals that the maximum value that x_2 can attain is $3\frac{1}{2}$. This implies that the labor constraint is non-binding but the capital constraint is binding. Given $\lambda_1 = 0$ and $\lambda_2 > 0$ then from (2) we obtain that $\lambda_2 = 5$. Substituting $(\lambda_1, \lambda_2) = (0, 5)$ into (3) reveals that the Kuhn-Tucker conditions are not satisfied.

Alternative 3

This alternative assumes $x_1, x_2 = 0$ and $x_3 > 0$. Observation of the labor and capital constraints reveals that the maximum value that x_3 can attain is $6\frac{2}{3}$. This implies that the labour constraint is binding but the capital constraint is not. Given $\lambda_2 = 0$ and $\lambda_1 > 0$ then from (3) we obtain that $\lambda_1 = 2\frac{2}{3}$. Substituting $(\lambda_1, \lambda_2) = (2\frac{2}{3}, 0)$ into (2) reveals that the Kuhn-Tucker conditions are not satisfied.

(b)

The dual to the maximization problem presented in (a) is as follows:

$$\min_{y_1, y_2} V = 20y_1 + 14y_2$$

subject to:

$$10y_1 + 6y_2 \geq 20$$
$$2y_1 + 4y_2 \geq 20$$
$$3y_1 + y_2 \geq 8$$
$$y_1, y_2, \geq 0$$

This problem can easily be solved graphically or by using the simplex method in linear programming. The graphical solution is provided in the following figure and is $(y_1, y_2) = (1.2, 4.4)$. Comparison with the solution in (a) reveals that the value of the Lagrangean multipliers and the dual variables are identical. This is true for the class of problems where both the objective function and constraints are linear.

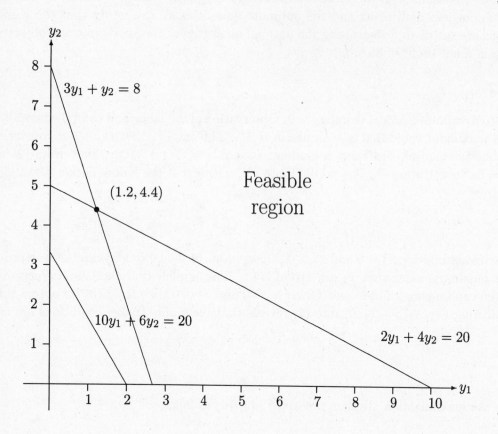

(c)

In general, the maximization problem with a linear objective function and constraints may be written as follows:

$$\max_{x_j} \sum_{j=1}^{n} c_j x_j$$

subject to:

$$\sum_{j=1}^{m} a_{ij} x_j \leq b_i \text{ for all } i = 1, \cdots, m$$

$$x_j \geq 0 \text{ for all } j = 1, \cdots, n$$

The corresponding Lagrangean may be written as follows:

$$L = \sum_{j=1}^{n} c_j x_j + \sum_{i=1}^{m} \lambda_i (b_i - \sum_{j=1}^{n} a_{ij} x_j)$$

The Kuhn-Tucker conditions are:

$\frac{\partial L}{\partial x_j} = c_j - \sum_{i=1}^{m} \lambda_i a_{ij} \leq 0; \ x_j \cdot \left(\frac{\partial L}{\partial x_j}\right) = 0$ for all j

$\frac{\partial L}{\partial \lambda_i} = b_i - \sum_{j=1}^{n} a_{ij} x_j \geq 0; \ \lambda_i \cdot \left(\frac{\partial L}{\partial \lambda_i}\right) = 0$ for all i

$x_j \geq 0$ for all j

$\lambda_i \geq 0$ for all i

The dual to the general maximization problem with a linear objective function and constraints is to **minimize** the following:

$$\min_{y_i} \sum_{i=1}^{m} b_i y_i$$

subject to:

$$\sum_{i=1}^{n} a_{ji} y_i \ \geq \ c_j \text{ for all } j = 1, \cdots, n$$

$$y_j \ \geq \ 0 \text{ for all } i = 1, \cdots, m$$

The objective in solving the dual problem is to minimize the *resource cost* or the quantity of the resources used multiplied by their shadow prices. The shadow prices or y_i measure the change in the **maximized** objective function in the primal problem from a one unit change in the ith resource.

The Lagrangean multipliers in the Kuhn-Tucker conditions may be interpreted as the **approximate** change in the maximized objective function from a one unit change in the constant of the ith constraining function. In the case of a linear objective function and constraints, the Lagrangean multiplier will measure the exact change in the objective function such that $\lambda_i = y_i$.

The problem illustrates that linear programming is a special case of nonlinear programming.

Quasiconcave Programming

Question 7

(a)

We will examine in turn the Arrow-Enthoven conditions listed at the beginning of the chapter.

1. For the defined maximization problem quasiconcavity of the objective function can be determined from the Hessian matrix of second-order partial derivatives.

$$H = \begin{bmatrix} \frac{\partial^2 y}{\partial x_1^2} & \frac{\partial^2 y}{\partial x_1 \cdot \partial x_2} \\ \frac{\partial^2 y}{\partial x_2 \cdot \partial x_1} & \frac{\partial^2 y}{\partial x_2^2} \end{bmatrix} = \begin{bmatrix} 2 & 0 \\ 0 & 2 \end{bmatrix}$$

where

$$|H_1| = 2 > 0, \quad |H| = 4 > 0$$

Thus, the leading principal minors of the Hessian are positive for all values of x_1 and x_2 such that the matrix is positive definite. This implies the objective function is strictly *convex* and, hence, the Arrow-Enthoven conditions are **not** satisfied.

2. $g_1(x_1, x_2)$ may be written as $k_1(x_1) + k_2(x_2)$ where $k_1' = 1 > 0$, $k_1'' = 0$ and $k_2' = 1 > 0$, $k_2'' = 0$. Thus, both k_1 and k_2 are convex and because the sum of convex functions is a convex function $g_1(x_1, x_2)$ must be convex. Because convexity implies quasiconvexity, $g_1(x_1, x_2)$ is quasiconvex. The $g_2(x_2)$ function is also linear which implies it is convex and, therefore, quasiconvex.

3. All the constraints are linear so the constraint qualification condition is automatically satisfied. Given that all the constraints are convex and differentiable the constraint qualification could also be shown to be satisfied by proving that there exists a vector $\underline{x}$ such that $g_i(x) < c_i$ where $\underline{x} \geq 0$. For example, the vector $(x_1, x_2) = (1, 1)$ satisfies this condition.

4. To verify the fourth condition is satisfied we need to evaluate at a potential solution $\underline{x}$ and confirm that **one** of the four conditions is satisfied.

(b)

Forming the Lagrangean yields:

$$L = (x_1 - 1)^2 + (x_2 - 2)^2 + \lambda_1(4 - x_1 - x_2) + \lambda_2(2 - x_2)$$

The Kuhn-Tucker conditions are:

(1) $\frac{\partial L}{\partial x_1} = 2x_1 - 2 - \lambda_1 \leq 0$; $x_1 \cdot \left(\frac{\partial L}{\partial x_1} \right) = 0$

(2) $\frac{\partial L}{\partial x_2} = 2x_2 - 4 - \lambda_1 - \lambda_2 \leq 0$; $x_2 \cdot \left(\frac{\partial L}{\partial x_2} \right) = 0$

(3) $\frac{\partial L}{\partial \lambda_1} = 4 - x_1 - x_2 \geq 0$; $\lambda_1 \cdot \left(\frac{\partial L}{\partial \lambda_1} \right) = 0$

(4) $\frac{\partial L}{\partial \lambda_2} = 2 - x_2 \geq 0$; $\lambda_2 \cdot \left(\frac{\partial L}{\partial \lambda_2} \right) = 0$

(5) $x_1 \geq 0$

(6) $x_2 \geq 0$

(7) $\lambda_1 \geq 0$

(8) $\lambda_2 \geq 0$

Substituting the solution $(x_1, x_2) = (4, 0)$ into the Kuhn-Tucker conditions we obtain from (1) that $\lambda_1 = 6$. Because the second constraint is not binding $\lambda_2 = 0$. Checking the Kuhn-Tucker conditions in turn:

(1) $0 \leq 0$ and $4 \cdot (0) = 0$

(2) $-10 \leq 0$ and $0 \cdot (-10) = 0$

(3) $0 \geq 0$ and $6 \cdot (0) = 0$

(4) $2 \geq 0$ and $0 \cdot (2) = 0$

(5) $4 \geq 0$

(6) $0 \geq 0$

(7) $6 \geq 0$

(8) $0 \geq 0$

Thus, the Kuhn-Tucker conditions are satisfied at the optimum $(x_1, x_2) = (4, 0)$ where the objective function achieves a value of 13.

(c)

Substituting $(x_1, x_2) = (1,2)$ into the Kuhn-Tucker conditions we may note that the first constraint is not binding such that $\lambda_1 = 0$.

(1) $0 \leq 0$ and $1 \cdot (0) = 0$
(2) $0 \leq 0$ and $2 \cdot (0) = 0$
(3) $1 \geq 0$ and $0 \cdot (1) = 0$
(4) $0 \geq 0$ and $0 \cdot (0) = 0$
(5) $1 \geq 0$
(6) $2 \geq 0$
(7) $0 \geq 0$
(8) $0 \geq 0$

Thus, the Kuhn-Tucker conditions are satisfied at $(x_1, x_2) = (1,2)$ and the objective function achieves a value of 0. We should observe that the Kuhn-Tucker conditions are only *necessary* conditions for a local maximum provided that the constraint qualification condition is satisfied and that $f(x)$ and all $g_i(x)$ are continuously differentiable. Because some value $\underline{x}$ satisfies the Kuhn-Tucker conditions does **not** necessarily mean it is the optimum. The only implication of a necessary condition is that the optimum must satisfy the condition but is does not preclude non-optimal values from satisfying the condition.

For problems where the Arrow-Enthoven conditions are satisfied then the Kuhn-Tucker conditions will be both *necessary* and *sufficient*. This means that the optimum must satisfy the conditions and that if an $\underline{x}$ does satisfy the conditions it must be an optimum. In other words, if $\underline{x}$ satisfies the Kuhn-Tucker conditions then we can be assured that that it is a unique optimum.

Necessary and Sufficient Conditions for a Maximum

Question 8

The general maximization problem may be written as follows:

$$\max_{x} \; f(x)$$

subject to:

$$
\begin{aligned}
cx &\leq b \\
x &\geq 0
\end{aligned}
$$

where c and $b > 0$.

The Kuhn-Tucker conditions are:

(1) $\frac{\partial L}{\partial x} = \frac{\partial f(x)}{\partial x} - \lambda c \leq 0$; $x \cdot \left(\frac{\partial L}{\partial x}\right) = 0$
(2) $\frac{\partial L}{\partial \lambda} = b - cx \geq 0$; $\lambda \cdot \left(\frac{\partial L}{\partial \lambda}\right) = 0$
(3) $x \geq 0$
(4) $\lambda \geq 0$

Defining x^* as the global minimum of $f(x)$ we may observe that if $0 < cx^* < b$ the Kuhn-Tucker conditions are satisfied. Because the constraint is not binding we may note that $\lambda = 0$ and given that $x^* > 0$ it must be the case that $\frac{\partial L}{\partial x} = 0$. The Kuhn-Tucker conditions are as follows:

(1) $\frac{\partial L}{\partial x} = \frac{\partial f(x)}{\partial x} = 0;\ x^* \cdot (0) = 0$

(2) $\frac{\partial L}{\partial \lambda} = b - cx^* > 0;\ 0 \cdot \left(\frac{\partial L}{\partial \lambda}\right) = 0$

(3) $x^* > 0$

(4) $\lambda = 0$

The problem demonstrates that the Kuhn-Tucker conditions along with the constraint quali-fication condition are *necessary* conditions for an optimum provided that $f(x)$ and all $g_i(x)$ are continuously differentiable. Satisfaction of the Kuhn-Tucker conditions by some $\underline{x}$ along with the constraint qualification does not, therefore, ensure that $\underline{x}$ is an optimum. If further conditions are imposed upon the problem such as the Arrow-Enthoven conditions or the concave programming criteria such that $f(x)$ is concave then the Kuhn-Tucker conditions will be both *necessary* and *sufficient*.

Chapter 9

Complex Numbers

Objectives

The questions in this chapter should help the reader master the fundamentals of complex numbers as used in economics. Readers who are able to answer all the questions should be able to:

1. Transform complex numbers in cartesian form to polar form (Question 1).

2. Transform complex numbers in exponential form to cartesian form (Question 2).

3. Apply *De Moivre's Theorem* (Question 3).

4. Add, subtract, and multiply complex numbers (Question 4).

5. Obtain the complex roots of a quadratic equation (Question 5).

Review

Rational numbers can be expressed as a ratio of integers. *Real numbers* include all rational and *irrational numbers* which cannot be expressed as a ratio of integers such as $\sqrt{2}$. Complex numbers contain a real and an imaginary part. Thus, real numbers are a subset of complex numbers. In general, the cartesian form of a complex number may be written as follows:

$$Z = a + bi$$

where a is a real number, bi is an imaginary number, and $i = \sqrt{-1}$.

The two complex numbers $(a \pm bi)$ are defined as the complex conjugate. In cartesian coordinates, the a represents the horizonal distance from the origin on the horizontal or real axis while bi represents the vertical distance b from the origin on the vertical or imaginary axis. The complex number, therefore, represents a point in this space.

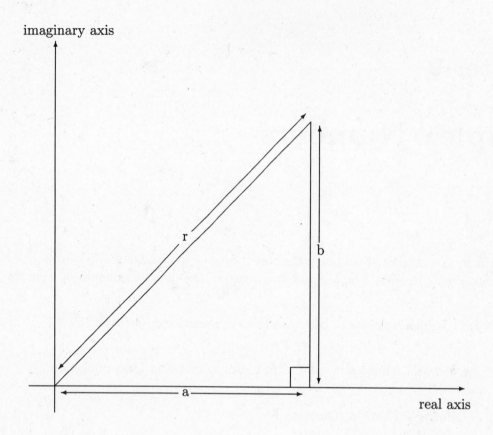

When using complex numbers in economics it is often convenient to transform them from the cartesian to polar and exponential forms and vice versa. The polar and exponential forms of the complex number defined above are given below:

$$\text{Polar Form} = \ r(\cos\theta \pm i\sin\theta)$$
$$\text{Exponential Form} = \ re^{\pm i\theta}$$

In transforming from the cartesian to polar form it is necessary to know that r, the modulus, is defined by $\sqrt{a^2 + b^2}$, and that

$$\frac{a}{r} \equiv \cos\theta$$
$$\frac{b}{r} \equiv \sin\theta$$

In working with the trigonometric functions "cos" and "sin" it is often more convenient to use radians rather than degrees. By noting that 360 deg. equals 2π we may obtain a correspondence for any angle θ in radians. In solving for θ from the cartesian form we restrict ourselves to the domain defined by 0 and 2π.

An important formula when working with complex numbers is *De Moivre's Theorem*:

$$[r(\cos\theta \pm i\sin\theta)]^n = r^n(\cos n\theta \pm i\sin n\theta)$$

which can be restated in cartesian and polar form as follows:

$$(a \pm bi)^n = [r(\cos\theta \pm i\sin\theta)]^n$$

It is sometimes necessary to add, subtract, and multiply complex numbers. Fortunately, complex numbers satisfy the usual laws of arithmetic.

$$
\begin{aligned}
(a + bi) + (c + di) &= (a + c) + (b + d)i \\
(a + bi) - (c + di) &= (a - c) + (b - d)i \\
(a + bi) \cdot (c + di) &= (ac - bd) + (ad + bc)i
\end{aligned}
$$

When solving higher order difference and differential equations it is often necessary to obtain the characteristic roots of the characteristic equation (see chapters 11 and 12). Solving the characteristic equation of linear second-order difference and differential equations with constant coefficients requires finding the roots to a quadratic equation of the following form.

$$x^2 + a_1 x + a_2 c = 0$$

There are three possible alternatives

1. $a_1^2 - 4a_2 > 0 \Rightarrow$ there are two distinct real roots which can be solved using the quadratic formula.

2. $a_1^2 - 4a_2 = 0 \Rightarrow$ there are two repeated roots defined as $\bar{x}_1 = \bar{x}_2 = -\frac{a_1}{2}$.

3. $a_1^2 - 4a_2 < 0 \Rightarrow$ there are two complex roots.

In the case where there are two complex roots, the roots to the quadratic equation are defined as follows:

$$\bar{x}_1, \bar{x}_2 = a \pm bi;$$

where

$$a = \frac{-a_1}{2}, \text{ and } b = \frac{\sqrt{4a_2 - a_1^2}}{2}$$

Further Reading

Readers requiring further references to the subject may consult Binmore [5] (chapter 12), Black and Bradley [7] (chapter 16), Bressler [10] (chapter 3), Chiang [13] (chapter 15 sections 15.2 and 15.3), Glaister [19] (chapter 14 section 14.3), and Sydsæter [31] (chapter 2). Sydsæter [31] (chapter 2), in particular, presents a rigorous and in depth discussion of complex numbers and their applications in economics.

Chapter 9 - Questions

Cartesian and Polar Form

Question 1

Write the following complex numbers in the polar form:

(i) $\sqrt{3} + 3i$

(ii) -1

(iii) $-2 - 2\sqrt{3}i$

(iv) $1 - i$

Exponential and Cartesian Form

Question 2

Write the following complex numbers in the cartesian form:

(i) $e^{-\pi i}$

(ii) $e^{\frac{\pi i}{2}}$

(iii) $2i + e^{2\pi i}$

(iv) $e^{\frac{\pi i}{4}}$

De Moivre's Theorem

Question 3

Use *De Moivre's Theorem* to find a and b if $(a + bi) = (\sqrt{2} - \sqrt{2}i)^4$.

Complex Algebra

Question 4

Solve the following:

(i) $(3 + 6i)(2 - i)$

(ii) $(\frac{1}{2} - i)^2$

(iii) $(3 + i) - (6 - 3i)$

(iv) $(2 + 3i)(3 - 4i) + (1 + i)^2$

Complex Roots

Question 5

Find the roots of the following quadratic equations:

(i) $2x^2 + x + 8 = 0$

(ii) $x^2 + x + \frac{1}{2} = 0$

Chapter 9 - Solutions

Cartesian and Polar Form

Question 1

(i)

First solve for r, the modulus,

$$r = \sqrt{(\sqrt{3})^2 + 3^2}$$

$$\Rightarrow r = \sqrt{12}$$

Next, solve for the angle θ and convert it into radians,

$$\cos\theta = \frac{\sqrt{3}}{\sqrt{12}} = 0.5 = 60\deg = \frac{\pi}{3}$$

$$\sin\theta = \frac{3}{\sqrt{12}} = 0.866 = 60\deg = \frac{\pi}{3}$$

In solving for θ, we must find the angle that satisfies **both** the $\cos\theta$ and $\sin\theta$ expressions simultaneously in the interval 0 to 2π. Combining the results we obtain the polar form of the complex number.

$$\sqrt{3} + 3i = \sqrt{12} \cdot (\cos(\frac{\pi}{3}) + i \cdot \sin(\frac{\pi}{3}))$$

(ii)

First obtain the modulus r,

$$r = \sqrt{(-1)^2} = 1$$

Next, obtain θ and convert into radians,

$$\cos\theta = -1 = 180\deg = \pi$$

$$\sin\theta = 0 = 180\deg = \pi$$

Combining the results,

$$-1 = \cos\pi + i \cdot \sin\pi$$

(iii)

First obtain the modulus, r

$$r = \sqrt{(-2)^2 + (-2\sqrt{3})^2}$$

$$\Rightarrow r = \sqrt{4 + 12}$$

$$\Rightarrow r = 4$$

Next, obtain θ and convert into radians,

$$\cos\theta = \frac{-2}{4} = 240\deg = \frac{4\pi}{3}$$

$$\sin\theta = \frac{-2\sqrt{3}}{4} = 240\deg = \frac{4\pi}{3}$$

Thus:

$$-2 - 2\cdot\sqrt{3}i = 4\cdot\left(\cos(\frac{4\pi}{3}) + i\cdot\sin(\frac{4\pi}{3})\right)$$

(iv)

First obtain the modulus, r,

$$r = \sqrt{1^2 + (-1)^2}$$

$$\Rightarrow r = \sqrt{2}$$

Next obtain θ and convert into radians,

$$\cos\theta = \frac{1}{\sqrt{2}} = \frac{\sqrt{2}}{2} = 315\deg = \frac{7\pi}{4}$$

$$\sin\theta = \frac{-1}{\sqrt{2}} = -\frac{\sqrt{2}}{2} = 315\deg = \frac{7\pi}{4}$$

Thus:

$$1 - i = \sqrt{2}\cdot\left(\cos(\frac{7\pi}{4}) + i\cdot\sin(\frac{7\pi}{4})\right)$$

Exponential and Cartesian Form

Question 2

(i)

From the definition of a complex number in exponential form we note that the r, the modulus, is defined as the constant term that multiplies the $e^{\pm i\theta}$ term. In this problem, therefore,

$$r = 1$$

The θ term in this problem is defined in radians as $-\pi = 180\,\deg$. The a and b terms in the cartesian form can be determined as follows:

$$a = r\cos\theta = 1 \cdot \cos 180\,\deg = -1$$

$$b = r\sin\theta = 1 \cdot \sin 180\,\deg = 0$$

Thus:

$$a + bi = -1$$

(ii)

Using the same approach as in (ii) we note that

$$r = 1$$

Similarly $\theta = \frac{\pi}{2} = 90\,\deg$ such that

$$a = r\cos\theta = 1 \cdot \cos 90\,\deg = 0$$

$$b = r\sin\theta = 1 \cdot \sin 90\,\deg = 1$$

Thus:

$$a + bi = i$$

(iii)

First, transform $e^{2\pi i}$ into cartesian form noting that $r = 1$ and $\theta = 2\pi = 360\,\deg$. The cartesian form is calculated as follows:

$$a = 1 \cdot \cos 360\,\deg = 1$$

$$b = 1 \cdot \sin 360\,\deg = 0$$

Thus:

$$1 \equiv e^{2\pi i}$$

Combining this result with $2i$ we obtain:

$a + bi = 1 + 2i.$

(iv)

Following the approach of the other problems we may note that $r = 1$ and $\theta = \frac{\pi}{4} = 45\,\text{deg}$. It follows, therefore,

$$a = 1 \cdot \cos 45\,\text{deg} = \frac{\sqrt{2}}{2}$$

$$b = 1 \cdot \sin 45\,\text{deg} = \frac{\sqrt{2}}{2}$$

Thus:

$$a + bi = \frac{\sqrt{2}}{2} + (\frac{\sqrt{2}}{2}) \cdot i$$

De Moivre's Theorem

Question 3

First we must transform $(\sqrt{2} - \sqrt{2}i)^4$ into polar form using *De Moivre's Theorem*. The modulus, r, is obtained as follows:

$$r = \sqrt{(\sqrt{2})^2 + (-\sqrt{2})^2} \Rightarrow r = 2$$

Next, obtain θ and then convert into radians

$$\cos\theta = \frac{\sqrt{2}}{2} = 315\,\text{deg} = \frac{7\pi}{4}$$

$$\sin\theta = \frac{-\sqrt{2}}{2} = 315\,\text{deg} = \frac{7\pi}{4}$$

Applying *De Moivre's Theorem*, the polar form is as follows:

$$2^4(\cos 4(\frac{7\pi}{4}) + i\sin 4(\frac{7\pi}{4})) = 16(\cos 7\pi + i\sin 7\pi)$$

The cartesian form can be obtained by observing that the modulus, r, is 16 and that because the trigonometric functions are periodic such they repeat themselves every 2π, then $\cos(7\pi) = \cos(\pi)$ and $\sin(7\pi) = \sin(\pi)$. It follows:

$$a = 16 \cdot \cos 180 \deg = 16 \cdot -1 = -16$$

$$b = 16 \cdot \sin 180 \deg = 16 \cdot 0 = 0$$

Thus:

$$(\sqrt{2} - \sqrt{2}i)^4 = -16$$

Complex Algebra

Question 4

(i)

Using the multiplication rule we can establish:

$$(3 + 6i)(2 - i) = (3 \cdot 2 - 6 \cdot (-1)) + (3 \cdot (-1) + 6 \cdot 2)i$$

$$\Rightarrow (3 + 6i)(2 - i) = 12 + 9i$$

(ii)

We first expand the terms,

$$(\frac{1}{2} - i)^2 = (\frac{1}{2} - i)(\frac{1}{2} - i)$$

We now apply the multiplication rule for complex numbers

$$(\frac{1}{2} - i)^2 = (\frac{1}{4} - 1) + (-\frac{1}{2} - \frac{1}{2})i$$

Summing the real numbers in the brackets we obtain

$$(\frac{1}{2} - i)^2 = -\frac{3}{4} - i$$

(iii)

Using the subtraction rule for complex numbers we obtain

$$(3 + i) - (6 - 3i) = (3 - 6) + (1 + 3)i$$

$$\Rightarrow (3 + 1) - (6 - 3i) = -3 + 4i$$

(iv)

This problem is best solved in parts. Using the multiplication rule for the first two terms we obtain

$$(2+3i)(3-4i) = (2 \cdot 3 + 3 \cdot 4) + (2 \cdot (-4) + 3 \cdot 3)i$$

$$\Rightarrow (2+3i)(3-4i) = 18 + i$$

Expanding the $(1+i)^2$ using the multiplication rule we obtain

$$(1+i)^2 = (1+i)(1+i) = (1-1) + (1+1)i$$

$$\Rightarrow (1+i)^2 = 2i$$

Collecting terms we obtain the result

$$(2+3i)(3-4i) + (1+i)^2 = (18+i) + (2i) = 18 + 3i$$

Complex Roots

Question 5

(i)

First, normalize the quadratic equation such that the coefficient on the x^2 term is equal to 1.

$$x^2 + \frac{1}{2}x + 4 = 0$$

Next, determine whether the roots are real or complex.

$$a_1^2 - 4a_2 = \frac{1}{4} - 16 = -15.75 < 0$$

There are, therefore, two *complex roots*. Using the formula to solve for the roots we obtain

$$a = -\frac{\frac{1}{2}}{2} = -\frac{1}{4}$$

$$b = \frac{1}{2} \cdot \sqrt{16 - \frac{1}{4}} = \frac{1}{2} \cdot \sqrt{15.75}$$

Thus:

$$\overline{x}_1, \overline{x}_2 = -0.25 \pm 0.5 \cdot \sqrt{15.75} \cdot i$$

(ii)

First we determine if the roots are real or complex.

$a_1^2 - 4a_2 = 1 - 2 = -1 < 0$
$\Rightarrow$ two complex roots

Applying the appropriate formula we obtain

$$a = \frac{-a_1}{2} = -\frac{1}{2}$$

$$b = \frac{1}{2} \cdot \sqrt{2 - 1} = \frac{1}{2}$$

Thus:

$\overline{x}_1, \overline{x}_2 = -\frac{1}{2} \pm \frac{1}{2} \cdot i.$

Chapter 10

Integration

Objectives

The questions in this chapter should help the reader master the basics of integration and appreciate its applications in business and economics. Readers who are able to answer all the questions in the chapter should be able to:

1. Solve indefinite integrals (Question 1).

2. Solve definite integrals (Questions 2 and 3).

3. Solve multiple integrals (Question 4).

4. Solve improper integrals (Question 5).

5. Apply the rules of integration to applied microeconomic problems (Questions 6, 7 and 8).

6. Apply the rules of integration to solve applied problems (Questions 9, 10 and 11).

Review

Integration is the method by which we derive a primitive function from a derived function. An understanding of integration is useful for the study of the time paths of economic variables, probability theory, and many other applications in economics.

There are a number of important rules in integration that will be applied in this chapter and are necessary for finding the primitive functions of variables from derived functions. The rules are perfectly general for all continuous functions where $f(x)$ is defined as the integrand and K is the constant of integration.

1. Multiplication by a constant rule:

$$\int cf(x)dx = c \int f(x)dx$$

where c is any real number

2. Power rule:

$$\int x^a dx = \frac{1}{a+1}x^{a+1} + K$$

where $a \neq -1$.

3. Exponential rule:

$$\int e^x dx = e^x + K$$

4. Logarithmic rule:

$$\int \frac{f'(x)}{f(x)} dx = \ln|f(x)| + K$$

where $f(x) \neq 0$.

5. Substitution rule:

$$\int f(u)\frac{du}{dx}dx = \int f(u)du$$

6. Integration by parts rule:

$$\int u(x)v'(x)dx = u(x)v(x) - \int v(x)u'(x)dx$$

These rules are defined in terms of *indefinite* integrals which are so called because the constant of integration is unknown.

Definite integrals have a numerical value and require an upper and lower limit of integration. Defining $F(x)$ as the primitive function of the derived function $f(x)$, a fundamental result of integration is the following:

$$\int_a^b f(x)dx = F(x)]_a^b = F(b) - F(a)$$

where b is the upper limit of integration and a is the lower limit of integration.

In addition to rules 1-6, when solving definite integrals the following results are useful to remember.

$$\int_a^b f(x)dx = -\int_b^a f(x)dx$$

$$\int_a^b f(x)dx = \int_a^c f(x) + \int_c^b f(x)dx$$

where $a < c < b$.

When integrating functions of several variables with respect to just one variable the same rules of integration apply and all other variables are treated as constants. When integrating functions of several variables with respect to each of the variables, then the problem is called a repeated or *multiple integral*. It is important to note that with repeated integrals changing the order of integration does not change the result. For problems that deal with areas of surfaces or volumes we often have to integrate for both width and height. An important result for such problems where $f(x, y)$ is a continuous function defined over the area $[a, b] \times [c, d]$ is:

$$\int_a^b (\int_c^d f(x,y)dy)dx = \int_c^d (\int_a^b f(x,y)dx)dy$$

The multiple integral can be interpreted in several ways including the volume of the ordinate set of $f(x, y)$ over $[a, b] \times [c, d]$, i.e., the volume beneath the surface defined by $f(x, y)$ and above the region $[a, b] \times [c, d]$.

Improper integrals arise when one of the limits of integration is infinite. The method of solution is to replace ∞ or $-\infty$ and then determine whether the limit exists as the limit of integration tends to ∞ or $-\infty$. For example:

$$\int_a^\infty f(x)dx \equiv \lim_{b \to \infty} \int_a^b f(x)dx$$

$$\int_{-\infty}^b f(x)dx \equiv \lim_{a \to -\infty} \int_a^b f(x)dx$$

If the limit exists then the integral is convergent, and if it does not, the integral is meaningless and is said to be divergent. An improper integral can also be defined when both the upper and lower limits of integration are real numbers if the value of the integrand is infinite at some point in between. For example if $a < 0 < b$ and the function $f(x)$ is infinite at 0 then the improper integral may be defined as follows:

$$\begin{aligned}
\int_a^b f(x)dx &= \int_a^0 f(x)dx + \int_0^b f(x)dx \\
&= \lim_{c \to 0-} \int_a^c f(x)dx + \lim_{c \to 0+} \int_c^b f(x)dx
\end{aligned}$$

Further Reading

This chapter provides only an introduction to the topic of integration. Interested readers are encouraged to practice their techniques with additional problems from other texts. We particularly recommend Haeussler and Paul [21] (chapters 16 and 17) for its detailed examples and applications and Dowling [17] (chapters 16 and 17) for its many worked examples. A good introduction to integral calculus for economists is also given in Holden and Pearson [23] (chapter 6), Chiang [13] (chapter 13), Ostrosky and Koch [25] (chapter 7). Sydsæter and Hammond [32] (chapters 10 and 11), in particular, provide very good examples of how to apply the methods of integration to solve economic problems.

A more detailed treatment of integration written for economists is provided by Binmore [5] (chapters 9 and 10) and Sydsæter [31] (chapter 4). Both of these references provide a number of exercises that should challenge most readers.

Chapter 10 - Questions

Indefinite Integrals

Question 1

Evaluate the following indefinite integrals,

(i)
$$\int \frac{x+2}{x+3}dx$$

(ii)
$$\int xe^{-x^2}dx$$

(iii)
$$\int \frac{1}{x-a}dx$$

(iv)
$$\int (x^4+3)^8 \cdot 9x^3 dx$$

(v)
$$\int (2x+5)(x^3-5x)dx$$

(vi)
$$\int x^3 \ln(x)dx$$

Definite Integrals

Question 2

Evaluate the following definite integrals

(i)
$$\int_0^4 (3x+5)dx$$

(ii)
$$\int_1^9 (x+3)(x^2-6)dx$$

Question 3

Suppose investment $(I(t))$ in an economy equals the change in the capital stock $(\frac{dK(t)}{dt})$ and is defined as follows:

$$I(t) = 3t^2$$

(i)

If the capital stock at period $t = 0$ is 10, what is the capital stock at period $t = 3$?

(ii)

Find the total **change** in the capital stock over the period $t = 5$ to $t = 7$.

Multiple Integrals

Question 4

Evaluate the following multiple integrals.

(i)

$$\int_0^2 \int_{\frac{y}{2}}^{y+2} dy dx$$

(ii)

$$\int \int x^2 + \frac{y}{x} dx dy$$

Improper Integrals

Question 5

Evaluate the following improper integrals

(i)

$$\int_2^\infty ce^{-x} dx$$

(ii)

$$\int_{-1}^4 x^{-2} dx$$

Cost Functions

Question 6

A firm's MC (marginal cost) function is:

$$MC = 4 + 6q + 30q^2$$

Find the TC (total cost) function, if TFC (fixed costs) are 100.

Consumer and Producer Surplus

Question 7

If the inverse demand function for a good q is defined as follows:

$$p = 10 - q - q^2$$

and the inverse supply function is:

$$p = q + 2$$

Find the consumer's and producer's surplus at the market equilibrium.

Question 8

For the following demand function

$$q = p^{\frac{-1}{\alpha}}$$

given that $\alpha > 0$, determine the values of α where the consumer's surplus is well defined.

Applications

Question 9

A professor takes a position at a university at the age of 30 and plans to retire at the age of 65. The average annual salary of professors at the university may be approximated by the following function:

$$S = 2,000y - \frac{100,000}{y} \quad \text{where } y \leq 65$$

where y is the age of a professor.

(a) Assuming the professor receives the average salary, how much would she receive as a starting and retiring salary?

(b) If the professor saves 10% of her gross salary until retiring and receives a zero real rate of interest on her savings, how much would she have saved in real terms upon retiring?

Question 10

The probability of waiting for treatment for a minor injury at the accident and emergency ward of a busy hospital is given by the following probability density function:

$$W = \frac{h^2}{9} \text{ for } 0 \le h \le 3.$$

where h is the time in hours.

What is the probability of waiting between 2 and 3 hours? and less than one hour?

Question 11

The monthly income distribution function of a community where the lowest income earner receives $100 and the highest income earner receives $20,000 is defined below.

$$f(m) = \frac{c}{m^2}$$

where $c > 0$ and m is monthly income.

(a) Determine the number of persons with an income less than $1,000/month and the number of persons with an income over $10,000/month. What is the ratio of "low" income earners to "high" income earners?

(b) What is the average monthly income per person?

Chapter 10 - Solutions

Indefinite Integrals

Question 1

(i)

First, transform the problem so that we can use the appropriate rules of integration. It can be verified that the problem may be written as follows:

$$\int (1 - \frac{1}{x+3})dx$$

Next, apply the power rule of integration to the first term and the logarithmic rule of integration to the second term and then add a constant of integration.

$$\int (1 - \frac{1}{x+3})dx = x - \ln(x+3) + K$$

(ii)

The easiest way to solve this problem is to use the substitution rule of integration. If we define $u = -x^2$ such that $\frac{du}{dx} = -2x$ then we can write $x = -\frac{1}{2} \cdot \frac{du}{dx}$. Performing the necessary substitutions we obtain:

$$\int -\frac{1}{2}e^u \cdot \frac{du}{dx}dx \Rightarrow \int -\frac{1}{2}e^u du$$

Using the multiplication by a constant rule and exponential rule of integration and adding a constant of integration we obtain the following:

$$-\frac{1}{2}e^u + K$$

Substituting back the value of $u = -x^2$ we obtain,

$$\int xe^{-x^2}dx = -\frac{1}{2}e^{-x^2} + K$$

This result can be checked by taking the derivative

$$\frac{d(-\frac{1}{2}e^{-x^2})}{dx} = -\frac{1}{2} \cdot \frac{d(e^{-x^2})}{dx}$$

$$= -\frac{1}{2} \cdot (-2xe^{-x^2})$$

$$= xe^{-x^2}$$

(iii)

The easiest way to solve the problem is to apply the substitution rule of integration. If we define $u = x - a$ such that $\frac{du}{dx} = 1$ and make the necessary substitutions we obtain:

$$\int \frac{1}{u} \cdot \frac{du}{dx} dx = \int \frac{1}{u} du$$

Applying the logarithmic rule of integration and adding a constant of integration we obtain:

$$= \ln(u) + K$$

Substituting back the value of $u = x - a$ we obtain:

$$\int \frac{1}{x - a} dx = \ln(x - a) + K$$

(iv)

The easiest way to solve this problem is to apply the substitution rule of integration. Define $u = x^4 + 3$ such that $\frac{du}{dx} = 4x^3$ and $\frac{9}{4} \cdot \frac{du}{dx} = 9x^3$. Making the necessary substitutions we obtain:

$$\int u^8 \cdot \frac{9}{4} \cdot \frac{du}{dx} dx = \int u^8 \cdot \frac{9}{4} du$$

Applying the multiplication by a constant rule of integration,

$$\int (x^4 + 3)^8 \cdot 9x^3 dx = \frac{9}{4} \cdot \int u^8 du$$

Applying the power rule of integration and adding a constant of integration

$$\int (x^4 + 3)^8 \cdot 9x^3 dx = \frac{9}{4} \cdot \frac{u^9}{9} + K$$

$$\int (x^4 + 3)^8 \cdot 9x^3 dx = \frac{1}{4} u^9 + K$$

Substituting the value of $u = x^4 + 3$ back into the solution we obtain:

$$\int (x^4 + 3)^8 \cdot 9x^3 dx = \frac{1}{4}(x^4 + 3)^9 + K$$

This result can be checked by taking the derivative,

$$\frac{d(\frac{1}{4}(x^4 + 3)^9)}{dx} = \frac{1}{4} \cdot \frac{d((x^4 + 3)^9)}{dx}$$

Applying the chain rule of differentiation, and letting $u = x^4 + 3$:

$$\frac{d(\frac{1}{4}(x^4 + 3)^9)}{dx} = \frac{1}{4}[9 \cdot \frac{du}{dx} \cdot u^8]$$

$$= 9x^3(x^4 + 3)^8$$

(v)

This problem is best solved using the integration by parts rule. First we must suitably define the functions $u(x)$ and $v(x)$ as follows:

$$u(x) = 2x + 5$$

$$v'(x) = x^3 - 5x$$

Applying the rules of differentiation and the power rule of integration it follows:

$$\frac{du}{dx} = 2$$

$$\int v'(x) = v(x) = \frac{1}{4}x^4 - \frac{5}{2}x^2$$

Applying the integration by parts rule

$$\int (2x + 5)(x^3 - 5x)dx = (2x + 5)(\frac{1}{4}x^4 - \frac{5}{2}x^2) - \int (\frac{1}{2}x^4 - 5x^2)dx$$

Integrating the second term of the RHS of the above expression using the power rule of integration we find that

$$\int (\frac{1}{2}x^4 - 5x^2)dx = \frac{x^5}{10} - \frac{5}{3}x^3 + K$$

substituting back into the expression we obtain:

$$\int (2x + 5)(x^3 - 5x)dx = (2x + 5)(\frac{1}{4}x^4 - \frac{5}{2}x^2) - \frac{x^5}{10} + \frac{5}{3}x^3 + K$$

Expanding terms we obtain:

$$\frac{1}{2}x^5 - 5x^3 + \frac{5}{4}x^4 - \frac{25}{2}x^2 - \frac{1}{10}x^5 + \frac{5}{3}x^3 + K$$

Collecting terms we obtain the primitive function

$$\frac{2}{5}x^5 + \frac{5}{4}x^4 - \frac{10}{3}x^3 - \frac{25}{2}x^2 + K$$

The result can be checked by differentiating the primitive functions and then factorizing.

(vi)

This problem can be solved using the integration by parts rule providing that the $u(x)$ and $v(x)$ functions are suitably defined as follows:

$$u(x) = \ln(x)$$

$$v'(x) = x^3$$

Thus,

$$\frac{du}{dx} = \frac{1}{x}$$

$$\int v'(x) = v(x) = \frac{x^4}{4}$$

Applying the integration by parts rule we obtain:

$$\int x^3 \ln(x)dx = \ln(x)\frac{x^4}{4} - \int \frac{x^3}{4}dx$$

Applying the multiplication by a constant rule and power rule of integration to the RHS term we obtain

$$\int x^3 \ln(x)dx = \ln(x)\frac{x^4}{4} - \frac{1}{4}\frac{x^4}{4} + K$$

Thus the solution is

$$\int x^3 \ln(x)dx = \ln(x)\frac{x^4}{4} - \frac{x^4}{16} + K$$

Definite Integrals

Question 2

(i)

The solution involves finding the primitive function and evaluating it at the upper limit (4) of integration and substracting the value of the primitive function at the lower limt (0) of integration. Applying the power rule of integration we obtain:

$$\int_0^4 (3x + 5)dx = (\frac{3}{2}x^2 + 5x)]_0^4$$

Evaluating at the upper and lower limits of integration yields

$$(\frac{3}{2}x^2 + 5x)]_0^4 = \frac{3}{2} \cdot 16 + 20 - 0$$

such that

$$\int_0^4 (3x + 5)dx = 44$$

(ii)

Expanding terms we obtain the following:

$$\int_1^9 (x + 3)(x^2 - 6)dx = \int_1^9 (x^3 + 3x^2 - 6x - 18)dx$$

Using the power rule of integration we may obtain the primitive function to be evaluated at the upper (9) and lower (1) limits of integration.

$$= (\frac{1}{4}x^4 + x^3 - 3x^2 - 18x)]_1^9$$

$$= (1640\frac{1}{4} + 729 - 243 - 162) - (\frac{1}{4} + 1 - 3 - 18)$$

$$= 1964\frac{1}{4} + 19\frac{3}{4}$$

such that,

$$\int_1^9 (x + 3)(x^2 - 6)dx = 1984$$

Question 3

(i)

Integrating the expression for $I(t)$ will give us an expression for the capital stock $K(t)$.

$$K(t) = \int I(t)dt = t^3 + K$$

where we are informed that $K = 10$. Thus, if $t = 3$ the capital stock is:

$$K(3) = 3^3 + 10 = 37$$

(ii)

The total change in the capital stock is obtained from the definite integral over the period $t = 5$ to $t = 7$, i.e.,

$$K(7) - K(5) = (7^3 + 10) - (5^3 + 10) = 218$$

Multiple Integrals

Question 4

(i)

First, we will integrate with respect to the variable y treating x as a constant and then we will integrate with respect to the variable x treating y as a constant. Which variable we choose first is not important as the order of integration does not change the result. Applying the power rule of integration we obtain:

$$\int_0^2 dx (y)]_{\frac{y}{2}}^{y+2}$$

Evaluating at the upper and lower limits of integration we obtain

$$\int_0^2 (y + 2 - \frac{y}{2}) dx$$

Simplifying the expression in brackets

$$\int_0^2 (\frac{y}{2} + 2) dx$$

Integrating with respect to the variable x

$$\int_0^2 (\frac{y}{2} + 2) dx = (x\frac{y}{2} + 2x)]_0^2$$

Evaluating at the upper and lower limits of integration we obtain:

$$\int_0^2 \int_{\frac{y}{2}}^{y+2} dy dx = y + 4$$

(ii)

In this multiple integral we integrate first with respect to x treating y as a constant and then integrate with respect to y treating x as a constant. The order of integration, however, does not change the result. In the first integration using the power rule and logarithmic rule we obtain the following:

$$\int (\int x^2 + \frac{y}{x} dx) dy = \int (\frac{1}{3}x^3 + y \ln(x) + K) dy$$

Performing the second integration using the power rule we obtain

$$\int \int x^2 + \frac{y}{x} dx dy = \frac{1}{3}x^3 y + \frac{1}{2}y^2 \ln(x) + K$$

where K is a constant of integration that is a function of both x and y.

Improper Integrals

Question 5

(i)

First, we obtain an equivalent problem defining b as the upper limit of integration and evaluating the integral as b tends to ∞.

$$\int_2^\infty ce^{-x}dx \equiv \lim_{b\to\infty} \int_2^b ce^{-x}dx$$

Using the exponential rule and multiplication by a constant rule of integration we obtain:

$$\lim_{b\to\infty} \int_2^b ce^{-x}dx = \lim_{b\to\infty} (-ce^{-x})]_2^b$$

Evaluating at the upper and lower limits of integration,

$$\lim_{b\to\infty} \int_2^\infty ce^{-x}dx = \lim_{b\to\infty} -ce^{-b} + ce^{-2}$$

As b tends to ∞ we may observe that $-ce^{-b}$ tends to 0. Thus:

$$\int_2^\infty ce^{-x}dx = ce^{-2}$$

(ii)

First, we observe that as x approaches 0 from the left the integrand tends to ∞. To evaluate the integral we must separate it into a sum of two integrals as follows:

$$\int_{-1}^4 x^{-2}dx = \int_{-1}^0 x^{-2}dx + \int_0^4 x^{-2}dx$$

Next, we will evaluate the two integrals separately. Substituting b in the upper limit of integration for the first integral and evaluating as b approaches 0 from the left we obtain the following:

$$\lim_{b\to 0-} \int_{-1}^b x^{-2} = \lim_{b\to 0-} (\frac{-1}{x})]_{-1}^b$$

Evaluating at the upper and lower limits of integration

$$\lim_{b\to 0-} (1 - \frac{1}{b})$$

As b tends to 0 then $\frac{1}{b}$ tends to ∞ and $1 - \frac{1}{b}$ tends to $-\infty$. The integral, therefore, does not exist and is divergent.

Cost Functions

Question 6

The total variable cost is the sum of the area beneath the marginal cost curve. It follows that:

$$\int \text{MC}dq = \text{TVC}$$

Integrating the MC for the specific problem using the power rule of integration we obtain:

$$\int (4 + 6q + 30q^2)dq = 4q + 3q^2 + 10q^3 + K$$

By definition, variable costs do not have a fixed element, such that $K = 0$. Thus:

$$\text{TVC} = 4q + 3q^2 + 10q^3$$

given that TFC are 100, TC is as follows:

$$\text{TC} = \text{TVC} + \text{TFC} = 100 + 4q + 3q^2 + 10q^3$$

Consumer and Producer Surplus

Question 7

At equilibrium, demand equals supply such that:

$$10 - q - q^2 = q + 2$$

$$8 - 2q - q^2 = 0$$

$$q^2 + 2q - 8 = 0$$

$$(q + 4)(q - 2) = 0$$

Because q cannot be negative then the equilibrium quantity is $q^* = 2$. Substituting q^* into either the inverse demand or supply functions we obtain the equilibrium price:

$$p^* = 10 - 2 - 4 = 4$$

Next, we must determine the consumer's surplus (CS) which is defined as the area above the market price and below the demand curve:

$$\text{CS} = \int_0^{q^*} p(q)dq - p^*q^*$$

Substituting in the specified inverse demand function and using the power rule of integration we obtain:

$$CS = (10q - \frac{q^2}{2} - \frac{q^3}{3})]_0^2 - p^*q^*$$

Evaluating at the limits of integration and noting that $p^*q^* = 8$ we obtain:

$$CS = \frac{22}{3}$$

The producer's surplus (PS) is defined as the area above the supply curve and below the market price:

$$PS = p^*q^* - \int_0^{q^*} p(q)dq$$

Substituting in the inverse supply function we obtain:

$$PS = p^*q^* - \int_0^2 (q + 2)dq$$

Using the power rule of integration:

$$PS = p^*q^* - (\frac{1}{2}q^2 + 2q)]_0^2$$

Thus, evaluating at the upper and lower limits of integration:

$$PS = 8 - 6 = 2$$

The producer and consumer surplus are illustrated below.

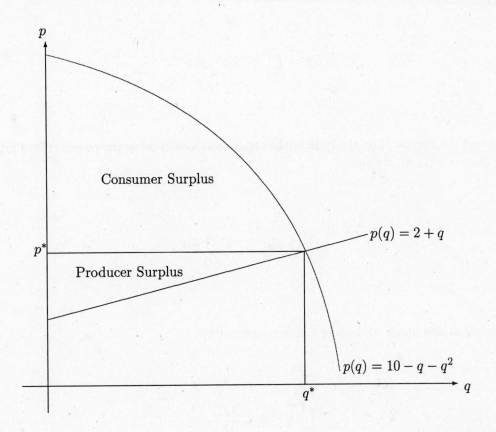

Question 8

The consumer's surplus is defined as the area above the market price and below the demand curve:

$$\text{CS} = \int_0^{q^*} p(q)dq - p^*q^*$$

where $p(q)$ is the inverse demand function and p^* and q^* are, respectively, the market price and quantity.

The inverse demand function for the given demand function is found as follows:

$$q = p^{\frac{-1}{\alpha}}$$

$$\Rightarrow q^{-1} = p^{\frac{1}{\alpha}}$$

$$\Rightarrow q^{-\alpha} = p$$

Substituting the inverse demand into the expression for CS

$$CS = \int_0^{q^*} q^{-\alpha} dq - p^* q^*$$

Integrating the first term on the RHS of the expression for CS using the power rule of integration (provided that $\alpha \neq 1$) we obtain:

$$\int_0^{q^*} q^{-\alpha} dq = (\frac{q^{1-\alpha}}{1-\alpha})]_0^{q^*}$$

Evaluating at the upper and lower limits of integration

$$\int_0^{q^*} q^{-\alpha} dq = \frac{(q^*)^{1-\alpha}}{1-\alpha}$$

We may observe that $p^* q^* = (q^*)^{1-\alpha}$. Thus provided that $\alpha < 1$,

$$CS = \frac{(q^*)^{1-\alpha}}{1-\alpha} - (q^*)^{1-\alpha}$$

which simplifies to the following:

$$CS = \frac{\alpha (q^*)^{1-\alpha}}{1-\alpha}$$

Thus, CS is well defined only if $\alpha < 1$. A graph of the inverse demand curve is given:

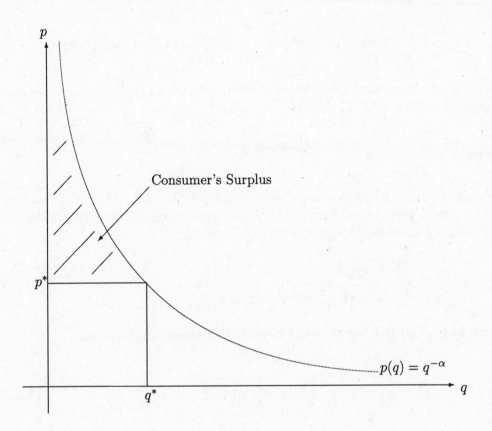

Applications

Question 9

(a)

The salary of the professor can be found by substituting her age into the function defined above as follows:

$$S_{30} = 2000 \cdot 30 - \frac{100,000}{30} = 56,666$$

$$S_{64} = 2000 \cdot 64 - \frac{100,000}{64} = 126,438$$

Because she retires on her 65th birthday her retiring salary is the salary she earns in her last year of employment when she is 64 years of age.

(b)

The professor's savings in real terms are defined by the following definite integral:

$$\int_{30}^{65} \frac{S}{10} dy = \int_{30}^{65} (200y - \frac{10,000}{y}) dy$$

Using the power rule and logarithmic rules of integration we obtain:

$$\int_{30}^{65} (2,000y - \frac{10,000}{y})dy = (1,000y^2 - 10,000\ln(y))]_{30}^{65}$$

Evaluating at the upper and lower limits of integration we find that her accumulated savings upon retirement are \$324,768.

Question 10

The probability of waiting between a and b hours is the definite integral of the probability density function where a and b are, respectively, the lower and upper limits of integration. The probability of waiting between 2 and 3 hours is calculated as follows:

$$\int_2^3 \frac{h^2}{9}dh = (\frac{h^3}{27})]_2^3 = \frac{19}{27}$$

The probability of waiting less than one hour is calculated in the same way.

$$\int_0^1 \frac{h^2}{9}dh = (\frac{h^3}{27})]_0^1 = \frac{1}{27}$$

Question 11

(a)

The number of people with an income between a and b \$/month is the definite integral of the income distribution function where a and b are, respectively, the lower and upper limits of integration. The number of people with an income less than \$1,000 a month is calculated as follows:

$$\int_{100}^{1,000} \frac{c}{m^2}dm = (-\frac{c}{m})]_{100}^{1,000} = -\frac{c}{1,000} + \frac{c}{100} = \frac{9c}{1,000}$$

The number of people with an income between \$10,000 and \$20,000 is calculated in a similar manner.

$$\int_{10,000}^{20,000} \frac{c}{m^2}dm = (-\frac{c}{m})]_{10,000}^{20,000} = -\frac{c}{20,000} + \frac{c}{10,000} = \frac{c}{20,000}$$

Multiplying both the denominator and numerator of the number of persons with an income less that \$1,000/month by 20 and dividing by the number of persons with an income between \$10,000 and \$20,000 we obtain the ratio 180.

(b)

To determine the average income in the community we must calculate the total income and then divide by the number of persons in the community. The total number of persons is found by integrating the income distribution function from the lowest to the highest income.

$$\int_{100}^{20,000} \frac{c}{m^2} dm = (-\frac{c}{m})]_{100}^{20,000} = -\frac{c}{20,000} + \frac{c}{100} = \frac{199c}{20,000}$$

The total income in the community is the definite integral of the income distribution function multiplied by m where the upper and lower limits of integration are, respectively, the highest and lowest levels of income.

$$\int_{100}^{20,000} m \cdot (\frac{c}{m^2}) dm = (c \ln(m))]_{100}^{20,000} = c(\ln(20,000) - \ln(100))$$

Performing the necessary calculations, the total income is some $5.30c$. Dividing total income by the total number of persons, the average income is calculated as follows:

$$\text{Average Income} \quad = \quad \frac{20,000 \cdot 5.298}{199} = \$532.49$$

Chapter 11

Difference Equations

Objectives

The questions in this chapter should help the reader solve linear difference equations and apply them in solving economic problems. Readers who can answer all the questions in the chapter should be able to:

1. Solve simple difference equations using the iterative method (Questions 1 and 2).

2. Solve linear first-order difference equations with constant coefficients and determine whether the time path is convergent and whether it oscillates (Questions 3 and 4).

3. Solve linear second-order difference equations with constant coefficients and determine whether the time path is convergent and whether it oscillates (Question 5).

4. Determine the stability of higher order linear difference equations (Question 6).

5. Solve higher order linear difference equations (Questions 7).

6. Solve a simultaneous system of difference equations (Question 8).

Review

Economic dynamics may be defined as the study of economic phenomena to past and future events (see Baumol [3]). Difference equations provide a step by step analysis of the time paths of economic variables. The approach is sometimes called period analysis as it is assumed that variables can change discontinuously from one period to the next. Difference equations are, therefore, appropriate for the analysis of dynamic problems in discrete time.

The first difference of a variable x in period t is defined as:

$$\Delta x_t = x_{t+1} - x_t$$

where Δ is called the difference operator and x_t is the first lag of x_{t+1}.

The second-order difference denoted by Δ^2 is the difference between successive first differences while the nth order difference is the successive difference between $n-1$ differences. A difference equation may be written in terms of difference operators or in terms of the lagged variables.

A difference equation may be of order *order* $1, \cdots, n$ where the order refers to the highest number of periods lagged. A difference equation is *homogeneous* if it contains only the lagged variables and no constant term and is *linear* if no variable is raised to power greater than one and is not multiplied by any other term of another period. A difference equation may also have constant coefficients on the lagged terms or the coefficients may vary with respect to time. An example of a linear, n^{th} order, non-homogeneous difference equation with constant coefficients is given below:

$$x_t + a_1 x_{t-1} + a_2 x_{t-2} + \cdots + a_n x_{t-n} = g(t)$$

where $a_n \neq 0$ and if we set $g(t) = 0$ we obtain the homogenous part of the difference equation.

A *solution* to a difference equation is an expression that is consistent with the original equation and contains no lagged variables. A *definite* solution to a difference equation is obtained when the initial condition or value of the variable in the first period is known. A *general* solution is obtained when the initial conditions are not specified.

For linear difference equations with constant coefficients there is a formula that may be applied to obtain the general solution. A linear first-order difference equation with constant coefficients may be defined as

$$x_t + a_1 x_{t-1} = c$$

where a_1 and c are constants.

The definite solution to such a difference equation is obtained by the following formula:

$$
\begin{aligned}
x_t &= \left(x_0 - \frac{c}{1 + a_1}\right) \cdot (-a_1)^t + \frac{c}{1 + a_1}, \quad \text{if } a_1 \neq -1 \\
x_t &= x_0 + ct, \quad \text{if } a_1 = -1
\end{aligned}
$$

In the case where the initial condition x_0 is not given then only a general solution is possible of the form:

$$x_t = A(-a_1)^t + \frac{c}{1 + a_1}$$

where $A(-a_1)^t$ is called the *complementary function* and is the solution to the homogeneous part of the difference equation and $\frac{c}{1+a_1}$ is the *particular solution*. The sum of the complementary function and particular solution always gives the solution to the difference equation.

The variable is said to be in a *stationary* state if it does not change with time. The time path of the variable is said to be *stable* if the variable converges to the particular solution over time. It follows, therefore, that if $|a_1| < 1$ the time path of the variable is convergent or stable. The time path of the variable is said to oscillate if $-a_1 < 0$ because the variable will be positive when t is even and negative when t is odd.

In the case of a linear second-order difference equations with constant coefficients, a formula may also be used to obtain a solution.

$$x_t + a_1 x_{t-1} + a_2 x_{t-2} = c$$

In this case, the particular solution defined as x_p is obtained as follows:

$$x_p = \frac{c}{1 + a_1 + a_2}, \quad \text{if } a_1 + a_2 \neq -1$$

$$x_p = (\frac{c}{2 + a_1})t, \quad \text{if } a_1 + a_2 = -1 \text{ and } a_1 \neq -2$$

$$x_p = (\frac{c}{2})t^2, \quad \text{if } a_1 + a_2 = -1 \text{ and } a_1 = -2$$

The complementary function, x_c, may be defined as:

$$x_c = A_1 r_1^t + A_2 r_2^t \text{ if } r_1 \neq r_2$$

$$x_c = A_1 r_1^t + A_2 t r_1^t \text{ if } r_1 = r_2$$

where r_1 and r_2 are the *characteristic roots* to the characteristic equation defined as the following quadratic equation:

$$r^2 + a_1 r + a_2 = 0$$

The characteristic roots may be real and unique, real and the same value or complex. The dominant root is the root with the largest absolute value. Where the roots are real, the time path of the variable is convergent if the absolute value of the dominant root is less than one. Where the roots are complex the time path is convergent if the modulus is less than one (see chapter 9 on complex numbers).

Convergence of an nth order difference equation can be determined from the n roots of the characteristic equation defined as follows:

$$r^n + a_1 r^{n-1} + \cdots + a_{n-1}r + a_n = 0$$

A neccessary and sufficient condition for convergence is that the determinants of the following n determinants are all positive. This is called *Schur's Theorem*.

$$\Delta_1 = \begin{vmatrix} 1 & a_n \\ a_n & 1 \end{vmatrix},$$

$$\Delta_2 = \begin{vmatrix} 1 & 0 & a_n & a_{n-1} \\ a_1 & 1 & 0 & a_n \\ a_n & 0 & 1 & a_1 \\ a_{n-1} & a_n & 0 & 1 \end{vmatrix}, \ldots$$

$$\Delta_n = \begin{vmatrix} 1 & 0 & \cdots & 0 & a_n & a_{n-1} & \cdots & a_1 \\ a_1 & 1 & \cdots & 0 & 0 & a_n & \cdots & a_2 \\ \cdots & \cdots & \cdots & \cdots & \cdots & \cdots & \cdots & \cdots \\ a_{n-1} & a_{n-2} & \cdots & 1 & 0 & 0 & \cdots & a_n \\ a_n & 0 & \cdots & 0 & 1 & a_1 & \cdots & a_{n-1} \\ a_{n-1} & a_n & \cdots & 0 & 0 & 1 & \cdots & a_{n-2} \\ \cdots & \cdots & \cdots & \cdots & \cdots & \cdots & \cdots & \cdots \\ a_1 & a_2 & \cdots & a_n & 0 & 0 & \cdots & 1 \end{vmatrix}$$

The ith determinant is constructed in four parts of dimension $i \times i$. The upper left area has a main diagonal of 1 with zeros above the diagonal and successively higher subscripts for the a_i

coefficients below the main diagonal. The upper right area has a main diagonal of a_n with zeros below the diagonal with successively smaller subscripts for a_i coefficients above the diagonal. The lower left and lower right areas are, respectively, the transpose of the upper right and upper left areas.

In the case where *all* the coefficients of the characteristic equation are positive a *sufficient* condition for stability is that:

$$1 > a_1 > a_2 > \cdots > a_n$$

In the case where not all the a_i coefficients are positive a *sufficient* condition for stability is that:

$$\sum_{i=1}^{n} |a_i| < 1$$

Further Reading

This chapter provides an introduction to difference equations and should enable the reader to solve a wide variety of economic problems. An introduction to difference equations with applications in economics is provided by Holden and Pearson [23] (chapter 10), Chiang (chapters 16-18), Dowling [17] (chapters 19 and 20) and Sydsæter and Hammond [32] (chapter 20).

Persons interested in solving more advanced problems are recommended to read Baumol [3], Goldberg [20] and Gandolfo [18]. Goldberg [20] is a classic reference in the social sciences on difference equations and Baumol [3] gives numerous applications of difference equations in economics. Gandolfo [18] provides an excellent treatment of linear difference equations and stability conditions.

Chapter 11 - Questions

Iterative Method

Question 1

Solve the following difference equations using the iterative method of solution.

(i) $y_{t+1} + a_1 y_t = c$ where $-1 < a_1 < 1$

(ii) $y_{t+1} + y_t = 0$

Question 2

Suppose that an individual invests a sum of money x_0 with a finance company and that the principal and interest that she owns in any period is 20 percent more than in the preceding period.

(i) Obtain an expression for the value of her investment in any period (x_t) in terms of the original deposit.

(ii) What is the value of x_0 that will ensure a stationary state for x_t?

Linear First-Order Difference Equations

Question 3

A closed economy is defined by the following equations:

$$
\begin{aligned}
c_t &= 100 + 0.9 y_{t-1} \\
\bar{I} &= 200
\end{aligned}
$$

where consumption in period t, lagged national income, and exogenous investment are respectively defined as c_t, y_{t-1}, and $\bar{I}$.

(a) Determine the equilibrium income for the economy

(b) Show whether the equilibrium level of income is stable and whether the time path oscillates.

Question 4

The demand for a commodity in period t, supply function in period $t+1$ and equlibrium condition are defined below where D_t, S_t, and P_t denote, respectively, the demand, supply and price of the good in period t.

(a) $D_t = 10 - P_t$

(b) $D_t = S_t$

(c) $S_{t+1} = -2 + 3P_t$

(a) What is the time path of the market price?

(b) Determine if there exists a market equilibrium and whether it is stable.

Stability

Question 5

Find the general solution to the following difference equations and show whether the system is stable or not.

(i) $x_t + 7x_{t-1} + 6x_{t-2} = 21$

(ii)

$$y_t = C_t + I_t$$

where

$$C_t = 10$$
$$I_t = 0.2y_{t-2}$$

(iii) $y_t = 2 - 2y_{t-2}$

Higher Order Linear Difference Equations

Question 6

The demand, supply and equilibrium conditions for a commodity are defined below where D_t, S_t, and p_t denote, respectively, the demand, supply and price of the good in period t and $d_1, s_1, s_2, s_3 > 0$ with $d_2 < 0$.

(a) $D_t = d_1 + d_2p_t$

(b) $D_t = S_t$

(c) $S_t = s_1p_{t-1} + s_2p_{t-2} + s_3p_{t-3}$

What are the necessary and sufficient conditions for the market to converge to an equilibrium?

Simultaneous System of Equations

Question 7

In a closed economy it is assumed that aggregate investment and consumption depend upon lagged values of national income as defined by the following equations.

$$c_t = 1,000 + 0.7y_{t-1} + 0.3y_{t-2}$$
$$I_t = 0.1(y_{t-1} - y_{t-2}) + 0.1(y_{t-2} - y_{t-3})$$

where c_t, I_t, are, respectively, aggregate consumption, aggregate investment in period t and y_{t-i} is national income in period $t - i$.

(a) Determine the equilibrium level of income for the economy.

(b) Solve for the time path of national income and determine whether it converges to the equilibrium level of income.

Question 8

Assume there are two economies where the exports of country one are the imports of country two and vice versa. The consumption, investment, imports and exports of each country are defined by the following set of equations.

Country One

$$
\begin{aligned}
C_{1t} &= 1,000 + 0.9y_{1t-1} \\
I_{1t} &= 100 \\
M_{1t} &= 100 + 0.1y_{1t-1} \\
X_{1t} &= M_{2t}
\end{aligned}
$$

Country Two

$$
\begin{aligned}
C_{2t} &= 2,000 + 0.8y_{2t-1} \\
I_{2t} &= 200 \\
M_{2t} &= 300 + 0.3y_{2t-1} \\
X_{2t} &= M_{1t}
\end{aligned}
$$

(a) Find the equilibrium level of national income in both countries.

(b) Find the time paths of national income for the two countries and determine if the dynamic system is stable.

Chapter 11 - Solutions

Iterative Method

Question 1

(i)

The iterative method of solution involves a step by step procedure to trace out the time path of the variable. The pattern that emerges can be used to induce the time path of the variable.

Rewriting the difference equation we obtain

$$y_{t+1} = c - a_1 y_t$$

Moving the equation forward one period,

$$y_{t+2} = c - a_1 y_{t+1}$$

Substituting in the expression for y_{t+1} we obtain:

$$y_{t+2} = c - a_1(c - a_1 y_t)$$

Collecting terms yields the following:

$$y_{t+2} = c - a_1 c + a_1^2 y_t$$

Repeating the process for the next time period we obtain :

$$y_{t+3} = c - a_1 y_{t+2}$$

substituting in the expression for y_{t+2} and collecting terms,

$$y_{t+3} = c - a_1 c + a_1^2 c - a_1^3 y_t$$

A pattern that emerges is that as $t \to \infty$ we obtain the sequence $c - a_1 c + a_1^2 c - a_1^3 c + a_1^4 c \cdots$ This is an infinite geometric series which has the following value provided that $-1 < a_1 < 1$.

$$\sum_{i=0}^{\infty} c(-1^i)(a_1)^i = \frac{c}{1 + a_1}$$

Defining A as an unknown constant, the pattern that emerges from the sequence is the general solution to the difference equation.

$$y_t = \frac{c}{1 + a_1} + (-a_1)^t A$$

Setting $t = 0$ we obtain the following:

$$y_0 = \frac{c}{1 + a_1} + A$$

$$\Rightarrow A = y_0 - \frac{c}{1 + a_1}$$

Substituting A back into the general solution we derive the definite solution to the difference equation.

$$y_t = \frac{c}{1 + a_1} + (-a_1)^t(y_0 - \frac{c}{1 + a_1})$$

This is identical to the general formula for the solution of a linear first-order difference equation with constant coefficients.

(ii)

The difference equation may be written as follows:

$$y_{t+1} = -y_t$$

Moving the equation forward one period and then substituting in the value for the previous period we obtain:

$$y_{t+2} = -y_{t+1} = -(-y_t) = y_t$$

Repeating the procedure,

$$y_{t+3} = -y_{t+2} = -(y_t) = -y_t$$

The general pattern or general solution that emerges is given below where A is an unknown constant.

$$y_t = (-1)^t \cdot A$$

If we set $t = 0$ then from the general solution we obtain:

$$y_0 = (-1)^0 \cdot A \Rightarrow y_0 = A$$

Substituting $y_0 = A$ back into the general expression for y_t we obtain the definite solution:

$$y_t = (-1)^t \cdot y_0$$

Question 2

(i)

The difference equation for this problem is

$$x_t = 1.2x_{t-1}$$

Thus, at $t = 1$

$$x_1 = 1.2x_0$$

and at $t = 2$

$$x_2 = 1.2x_1$$

If we substitute the expression for x_0 into x_2 we obtain

$$x_2 = 1.2^2 x_0$$

Successive iterations reveal that the solution to the difference equation is:

$$x_t = 1.2^t x_0$$

(ii)

Because $1.2 > 1$ as $t \to \infty$ then $x_t \to \infty$ if $x_0 > 0$. Thus, only if $x_0 = 0$ do we have a stationary state.

Linear First-Order Difference Equations

Question 3

(a)

First we will solve for the equilibrium level of income. From our knowledge of the income identity in macroeconomics we note the following:

$$y_t = c_t + \overline{I}$$

Substituting c_t and $\overline{I}$ from the equations that define the economy we obtain:

$$y_t = 100 + 0.9y_{t-1} + 200$$

$$\Rightarrow y_t = 300 + 0.9y_{t-1}$$

If the equilibrium level of income y_t is stationary it will be unchanging with respect to time such that:

$$y_t = y_{t-1} = y_e$$

Substituting y_e into the national income identity we obtain the particular solution to the difference equation.

$$y_e = 300 + 0.9y_e \Rightarrow y_e = 3,000$$

(b)

The time path of national income can be found by solving the following difference equation:

$$y_t = 300 + 0.9y_{t-1}$$

Using the general formula for a first-order linear difference equation with constant coefficients we may define $a_1 = -0.9$ and $c = 300$. Thus:

$$y_t = (y_0 - \frac{300}{1 - 0.9}) \cdot 0.9^t + \frac{300}{1 - 0.9}$$

$$= (y_0 - 3000) \cdot (0.9)^t + 3000$$

Stability can be assured because $|-0.9| < 1$. Convergence to $y_e = 3,000$ can also be deduced by noting that as $t \to \infty$, $y_t \to 3000$ irrespective of whether y_0 is greater than or less than y_e. Because $-(-0.9) > 0$ the time path is non-oscillating.

Question 4

(a)

At a market equilibrium, it must be the case that:

$$S_t = D_t$$

$$\Rightarrow -2 + 3P_{t-1} = 10 - P_t$$

$$\Rightarrow P_t + 3P_{t-1} = 12$$

$$\Rightarrow P_t = 12 - 3P_{t-1}$$

To solve for the time path of the price variable we need to solve the above first-order, linear difference equation with constant coefficients. Applying the general formula and noting that $a_1 = 3$ and $c = 12$ we obtain the definite solution.

$$P_t = (P_0 - \frac{12}{(1+3)}) \cdot (-3)^t + \frac{12}{(1+3)}$$

$$\Rightarrow P_t = (P_0 - 3) \cdot (-3)^t + 3$$

(b)

We may observe that the time path is *not* stable because $|3| > 0$. Substituting into the time path any initial value for P_0 other than $P_0 = 3$ will ensure that the next period's price will be even further away from P_0. The time path is, therefore, divergent. A market equilibrium only exists, therefore, when $P_0 = P_e = 3$.

Stability

Question 5

(i)

From the general form of second-order difference equations we note:

$$y_t + a_1 y_{t-1} + a_2 y_{t-2} = c$$

where, if $a_1 + a_2 \neq -1$, then the particular solution is given by:

$$y_p = \frac{c}{1 + a_1 + a_2}$$

For the given difference equation, we note that:

$$7 + 6 \neq -1$$

and so:

$$x_p = \frac{21}{14} = \frac{3}{2}$$

To determine the *complementary function* we note that if there are two distinct real roots or two complex roots the complementary function may be written as follows:

$$x_c = A_1 r_1^t + A_2 r_2^t$$

where r_1 and r_2 are the characteristic roots of the *characteristic equation*. In the case of two repeated real roots the complementary function has the following form:

$$x_c = A_1 r_1^t + A_2 t r_1^t$$

The characteristic equation is obtained from the homogeneous part of the difference equation as follows:

$$r^2 + 7r + 6 = 0$$

Because we have:

$$a_1^2 - 4a_2 = 49 - 24 = 25 > 0$$

$\bar{r}_1, \bar{r}_2$ are two *distinct real roots*.

Factorizing the characteristic equation we obtain

$$(r + 1)(r + 6) = 0$$

Thus $\bar{r}_1, \bar{r}_2 = -1$ and -6.

Substituting $\bar{r}_1, \bar{r}_2$ into the general expression for the *complementary function* we obtain:

$$x_c = A_1(-1)^t + A_2(-6)^t$$

Because $x_t = x_p + x_c$, the general solution is as follows:

$$x_t = \frac{3}{2} + A_1(-1)^t + A_2(-6)^t$$

Stability requires that the absolute value of the dominant root be less than 1. Because $|6| > 0$ the system is not stable and the variable does not converge to an intertemporal equilibrium.

In this case $|-6| > 1$, and so the system is **not** stable.

(ii)

Combining the expressions we obtain:

$$y_t = 10 + 0.2y_{t-2}$$

or in standard form to apply the general formula:

$$y_t - 0.2y_{t-2} = 10$$

We note that $-0.2 \neq -1$ and so the *particular solution* can be found as follows:

$$y_p = \frac{10}{1 - 0.2} = \frac{10}{0.8} = 12.5$$

If they are two distinct real roots or two complex roots the *complementary solution* may be written as:

$$y_c = A_1 r_1^t + A_2 r_2^t$$

where r_1 and r_2 are the characteristic roots of the *characteristic equation*. In the case of two repeated real roots the complementary function has the following form:

$$y_c = A_1 r_1^t + A_2 t r_1^t$$

The characteristic equation is obtained from the homogeneous part of the difference equation as follows:

$$r^2 - 0.2 = 0$$

Thus:

$$\bar{r}_1, \bar{r}_2 = \pm\sqrt{0.2}$$

Substituting this into the complementary function yields:

$$y_c = A_1(\sqrt{0.2})^t + A_2(-\sqrt{0.2})^t$$

Because: $y_t = y_p + y_c$, in this case we have:

$$y_t = 12.5 + A_1(\sqrt{0.2})^t + A_2(-\sqrt{0.2})^t$$

Stability is determined by the absolute value of the dominant root. Because $|\sqrt{0.2}| < 1$, as $t \to \infty$ y_t converges to 12.5 and is stable.

(iii)

Rewriting the difference equation in the standard form to apply the general formula:

$$y_t + 2y_{t-2} = 2$$

Since $2 \neq -1$, we note that the *particular solution* is found as follows:

$$y_p = \frac{21}{1+2} = \frac{21}{3} = 7$$

In the case of two distinct real roots or two complex roots the general form of the *complementary function* is as follows:

$$y_c = A_1 r_1^t + A_2 r_2^t$$

where r_1 and r_2 are the characteristic roots of the *characteristic equation.* In the case of two repeated real roots the complementary function has the following form:

$$x_c = A_1 r_1^t + A_2 t r_1^t$$

The characteristic equation is obtained from the homogeneous part of the difference equation as follows:

$$r^2 + 2 = 0$$

where,

$$a_1^2 - 4a_2 = 0 - 8 = -8 < 0$$

Thus, the characteristic equation has two complex roots which may written in cartesian form as follows:

$$x \pm yi$$

where:

$$x = -\frac{1}{2} a_1 = 0$$

$$y = \frac{\sqrt{4a_2 - a_1^2}}{2} = \frac{\sqrt{8}}{2} = \frac{2\sqrt{2}}{2} = \sqrt{2}$$

and so:

$$r_1, r_2 = \pm(\sqrt{2})i$$

It is conventional, however, to write the solution **without** imaginary parts. We must, therefore, transform the roots from cartesian to polar form (see chapter 9 - Complex Numbers).

From our knowlege of trigonometric functions for the complex conjugate $(a \pm bi)$.

$$\sin \theta = \frac{b}{r}$$

$$\cos \theta = \frac{a}{r}$$

where the modulus, r, is defined as

$$r = \sqrt{a^2 + b^2}$$

Thus we obtain the following

$$r = \sqrt{2}$$

and consequently:

$$\sin\theta = \frac{\sqrt{2}}{\sqrt{2}} = 1 = 90\deg = \frac{\pi}{2}$$

$$\cos\theta = \frac{0}{\sqrt{2}} = 0 = 90\deg = \frac{\pi}{2}$$

Thus, we may rewrite the complex roots in the *polar form*:

$$x + yi = \sqrt{2}\cdot(\cos\frac{\pi}{2} \pm i\sin\frac{\pi}{2})$$

Substituting the complex roots into the general expression for the complementary function yields:

$$y_c = A_1\cdot(\sqrt{2})^t\cdot(\cos\frac{\pi}{2} + i\sin\frac{\pi}{2})^t + A_2\cdot(\sqrt{2})^t\cdot(\cos\frac{\pi}{2} - i\sin\frac{\pi}{2})^t$$

From *De Moivre's Theorem*:

$$(\sqrt{2})^t\cdot(\cos\frac{\pi}{2} + i\sin\frac{\pi}{2})^t = (\sqrt{2})^t\cdot(\cos(\frac{\pi}{2})\cdot t + i\sin(\frac{\pi}{2})\cdot t)$$

Applying the theorem to the complementary function and collecting terms:

$$y_c = (\sqrt{2})^t((A_1 + A_2)\cdot\cos(\frac{\pi}{2})\cdot t + (A_1 - A_2)\cdot i\sin(\frac{\pi}{2})\cdot t)$$

Now, if we define:

$$B_1 = A_1 + A_2$$

$$B_2 = (A_1 - A_2)i$$

then we obtain:

$$y_c = (\sqrt{2})^t(B_1\cdot\cos(\frac{\pi}{2})\cdot t + B_2\sin(\frac{\pi}{2})\cdot t)$$

Since $y_t = y_p + y_c$, we have:

$$y_t = 7 + \sqrt{2}^t(B_1\cdot\cos(\frac{\pi}{2})t + B_2\sin(\frac{\pi}{2})t)$$

Stability of the system is determined if the modulus, r, is less than 1 in absolute value. In this case, we have $\sqrt{2} > 1$ and so the system is **not** stable.

Higher Order Linear Difference Equations

Question 6

Setting $D_t = S_t$ and collecting terms we obtain:

$$d_1 + d_2 p_t - s_1 p_{t-1} - s_2 p_{t-2} - s_3 p_{t-3} = 0$$

Stability requires that if p_t does not equal the equilibrium price, p_e, it will nevertheless converge to p_e as $t \to \infty$. To determine the stability conditions we need to obtain the characteristic equation and verify that the absolute value of all its roots are strictly less than one. The homogeneous part of the third order difference equation is given below.

$$p_t - \frac{s_1}{d_2} p_{t-1} - \frac{s_2}{d_2} p_{t-2} - \frac{s_3}{d_2} p_{t-3} = 0$$

The characteristic equation of the homogeneous part of the difference equation is defined as

$$r^3 - \frac{s_1}{d_2} r^2 - \frac{s_2}{d_2} r - \frac{s_3}{d_2} = 0$$

The necessary and sufficient conditions for stability are that the determinants of the following matrices are all positive (Schur's Theorem).

$$\Delta_1 = \begin{vmatrix} 1 & -\frac{s_3}{d_2} \\ -\frac{s_3}{d_2} & 1 \end{vmatrix},$$

$$\Delta_2 = \begin{vmatrix} 1 & 0 & -\frac{s_3}{d_2} & -\frac{s_2}{d_2} \\ -\frac{s_1}{d_2} & 1 & 0 & -\frac{s_3}{d_2} \\ -\frac{s_3}{d_2} & 0 & 1 & -\frac{s_1}{d_2} \\ -\frac{s_2}{d_2} & -\frac{s_3}{d_2} & 0 & 1 \end{vmatrix},$$

$$\Delta_3 = \begin{vmatrix} 1 & 0 & 0 & -\frac{s_3}{d_2} & -\frac{s_2}{d_2} & -\frac{s_1}{d_2} \\ -\frac{s_1}{d_2} & 1 & 0 & 0 & -\frac{s_3}{d_2} & -\frac{s_2}{d_2} \\ -\frac{s_2}{d_2} & -\frac{s_1}{d_2} & 1 & 0 & 0 & -\frac{s_3}{d_2} \\ -\frac{s_3}{d_2} & 0 & 0 & 1 & \frac{s_1}{d_2} & -\frac{s_2}{d_2} \\ -\frac{s_2}{d_2} & -\frac{s_3}{d_2} & 0 & 0 & 1 & -\frac{s_1}{d_2} \\ -\frac{s_1}{d_2} & -\frac{s_2}{d_2} & -\frac{s_3}{d_2} & 0 & 0 & 1 \end{vmatrix}$$

Simultaneous System of Equations

Question 7

(a)

Using the national income identity that $y_t = c_t + I_t$ we can substitute in the expressions for c_t and I_t to obtain the following third order difference equation.

$$y_t - 0.8 y_{t-1} - 0.3 y_{t-2} + 0.1 y_{t-3} = 1,000$$

If the economy has a stationary intertemporal equilibrium, national income will be unchanging with respect to time at the equilibrium. This requires that $y_t = y_{t-1} = y_{t-2} = y_{t-3} = y_e$. This particular solution, however, does not satisfy the difference equation. The intertemporal equilibrium must, therefore, change with respect to time. One potential equilibrium is that $y_t = At$ where A is an undetermined constant. Substituting this potential solution into the difference equation we obtain:

$$A(t) - 0.8A(t-1) - 0.3A(t-2) + 0.1A(t-3) = 1,000$$

$$\Rightarrow A = \frac{1,000}{(1 - 0.8 - 0.3 + 0.1)t + 0.8 + 0.6 - 0.3}$$

$$\Rightarrow A = \frac{1,000}{1.1}$$

$$\Rightarrow A = 909.09$$

The potential solution clearly satisfies the difference equation and the intertemporal equilibrium or particular solution is

$$y_e = At = 909.09t$$

(b)

To solve for the time path of national income we must derive the complementary function. The general form of the complementary function is as follows:

<u>Case One: Three Distinct Real Roots</u>

$$y_c = A_1 \overline{r_1}^t + A_2 \overline{r_2}^t + A_3 \overline{r_3}^t$$

<u>Case Two: Three Identical Real Roots</u>

$$y_c = A_1 \overline{r_1}^t + A_2 t \overline{r_1}^t + A_3 t^2 \overline{r_1}^t$$

<u>Case Three: One Real Root and Two Complex Roots</u>

$$y_c = A_1 \overline{r_1} + r^t (A_2 \cos \theta t + A_3 \sin \theta t)$$

where A_1, A_2, A_3 are unknown constants and r is the modulus for the complex roots.

The roots of the complementary function can be obtained from the chracteristic equation which is defined below.

$$r^3 - 0.8r^2 - 0.3r + 0.1 = 0$$

One way to obtain the value of the roots of a characteristic equation of order n is to guess the value of **one** of the roots and then verify if it satisfies the characteristic equation. The characteristic equation can then be written in terms of one of the known roots and a second characteristic equation of order $n - 1$. The process can then be repeated to solve out for the other roots.

As a first guess for one of the roots we will try all integers starting with 1 and verify if the guess is a characteristic root.

$$1^3 - 0.8(1^2) - 0.3 + 0.1 = 0$$

We are particularly fortunate that our first guess is a characteristic root. If we define the characteristic roots as $\overline{r_1}, \overline{r_2}, \overline{r_3}$ the characteristic equation can always be written as follows:

$$(r - \overline{r_1})(r - \overline{r_2})(r - \overline{r_3}) = 0$$

We have already found that $\overline{r_1} = 1$. The characteristic equation can, therefore, be split into a part containing $\overline{r_1}$ and a part that is of order $n - 1 = 2$.

$$(r - 1)(r^2 + a_1 r + a_2) = 0$$

The original form of the characteristic equation can be obtained by expansion of the factors. Given the original characteristic equation, multiplication of the above expression yields the following equations in terms of a_1 and a_2.

$$
\begin{aligned}
(a_2 - a_1)r &= -0.3r \\
(a_1 - 1)r^2 &= -0.8r^2 \\
-a_2 &= 0.1
\end{aligned}
$$

Solving the equations yields that $a_1 = 0.2$ and $a_2 = -0.1$. These are **not** the other two characteristic roots. However, using the quadratic formula we can solve for the other two roots as follows:

$$\overline{r_2}, \overline{r_3} = \frac{-a_1 \pm \sqrt{a_1^2 - 4a_2}}{2}$$

Thus, we obtain the following:

$$
\begin{aligned}
\overline{r_1} &= 1 \\
\overline{r_2} &= 0.231665 \\
\overline{r_3} &= -0.4316625
\end{aligned}
$$

There are three distinct real roots. Substituting into the general expression for the complementary function we obtain:

$$y_c = A_1 + A_2(0.231665)^t + A_3(-0.4316625)^t$$

Summing the complementary function and the particular solution yields the time path of national income and a general solution to the difference equation.

$$y_t = A_1 + A_2(0.231665)^t + A_3(-0.4316625)^t + 909.09t$$

A necessary and sufficient condition for national income to converge to an intertemporal equilibrium is that the absolute value of the roots of the characteristic equation are all strictly less than 1. It follows, therefore, that national income does **not** converge to the intertemporal equilibrium.

Question 8

(a)

First, we must obtain an expression for both y_{1t} and y_{2t}. Using the national income identity where $y_{it} = C_{it} + I_{it} + X_{it} - M_{it}$ and making the necessary substitutions we obtain

$$
\begin{aligned}
y_{1t} &= 1,300 + 0.8y_{1t-1} + 0.3y_{2t-1} \\
y_{2t} &= 2,000 + 0.5y_{2t-1} + 0.1y_{1t-1}
\end{aligned}
$$

One possible solution for an intertemporal equilibrium is that $y_{1t} = y_{1t-1} = y_{1e}$ and $y_{2t} = y_{2t-1} = y_{2e}$. To test this particular solution we must make the necessary substitutions and then solve for the two unknowns using the two equations. From the national income identity of country one we obtain:

$$y_{1e} - 0.8y_{1e} = 1,300 + 0.3y_{2e}$$

$$\Rightarrow 0.1y_{1e} = 650 + 0.15y_{2e}$$

Substituting into the income identity for country two and solving yields $y_{2e} = 7,571.43$. Substituting into the national income identity for country one yields $y_{1e} = 17,857.14$. This particular solution is consistent with the two difference equations.

(b)

To solve for the time path of national income for the two countries we need to obtain the complementary function. From the solution of single difference equations we may adopt the following possible solution for the complementary functions where r is the characteristic root.

$$y_{1t} = A_1 r^t$$

$$y_{2t} = A_2 r^t$$

The complementary function represents the solution to the homogeneous part of the difference equation. If we move the period forward one period such that $y_{1t} \to y_{1t+1}$ and $y_{1t-1} \to y_{1t}$ and likewise for y_{2t} and then substitute the possible solutions for y_{1t} and y_{2t} into the homogeneous part of the difference equations we obtain:

$$A_1 r^{t+1} = A_1 0.8 r^t + A_2 0.3 r^t$$
$$A_2 r^{t+1} = A_1 0.1 r^t + A_2 0.5 r^t$$

A general solution to the difference equation requires that we solve for the characteristic roots. Remembering that $r^t r = r^{t+1}$ we may write the system of equations in the following way.

$$r^t([0.8 - r]A_1 + A_2 0.3) = 0$$
$$r^t(A_1 0.1 + [0.5 - r]A_2) = 0$$

The trial solutions will only be a solution to the system of difference equations if the above conditions are satisfied. This requires that:

$$[0.8 - r]A_1 + A_2 0.3 = 0$$
$$A_1 0.1 + [0.5 - r]A_2 = 0$$

To obtain the characteristic equation we can exclude the trivial case where $A_1 = A_2 = 0$. Next, we should define the conditions in matrix form.

$$R = \begin{bmatrix} (0.8 - r) & 0.3 \\ 0.1 & (0.5 - r) \end{bmatrix},$$

$$A = \begin{bmatrix} A_1 \\ A_2 \end{bmatrix}$$

In matrix form the conditions may be written as follows:

$$RA = 0$$

The matrix R must be singular if there is to be a non-trivial solution to the set of conditions. From our knowledge of matrix algebra (see chapter 2) this requires that the determinant of the matrix R equal zero, i.e.,

$$|R| = \begin{vmatrix} (0.8 - r) & 0.3 \\ 0.1 & (0.5 - r) \end{vmatrix} = 0,$$

Expanding the determinant of the matrix R yields the charcteristic equation.

$$r^2 - 1.3r + 0.37 = 0$$

Solving for the roots of the characteristic equation using the quadratic formula we obtain the following.

$$\overline{r_1}, \overline{r_2} = \frac{1.3 \pm \sqrt{1.3^2 - 4 \cdot 0.37}}{2}$$

$$\Rightarrow \overline{r_1} = 0.8791288, \quad \overline{r_2} = 0.4208712$$

The two characteristic roots are real and distinct. It can be shown that both $\overline{r_1}$ and $\overline{r_2}$ satisfy the original set of conditions imposed by the trial solution. Consequently, there are two sets of solutions. In the case where $r = \overline{r_1}$ we may denote the solution by A_{11} and A_{21}.

$$[0.8 - \overline{r_1}]A_{11} + A_{21}0.3 \;=\; 0$$
$$A_{11}0.1 + [0.5 - \overline{r_1}]A_{21} \;=\; 0$$

With no loss in generality we can arbitrarily fix $A_{11} = 1$ and solve for A_{21} in terms of $\overline{r_1}$ and the coefficients and obtain the following.

$$A_{21} = \frac{0.1}{(\overline{r_1} - 0.5)} = \frac{\overline{r_1} - 0.8}{0.3}$$

Likewise if we substitute $\overline{r_2}$ into the set of conditions and define the solution by A_{12} and A_{22} and set $A_{12} = 1$ we obtain an expression for A_{22}.

$$A_{22} = \frac{0.1}{(\overline{r_2} - 0.5)} = \frac{\overline{r_2} - 0.8}{0.3}$$

A general solution for the time paths of national income must include the two solutions obtained from the two characteristic roots. If we define B_1 and B_2 as two unknown constants a general solution to the homogeneous part of the dynamic system is as follows:

$$y_{1t} \;=\; B_1 A_{11}\overline{r_1}^{\,t} + B_2 A_{12}\overline{r_2}^{\,t}$$
$$y_{1t} \;=\; B_1 A_{21}\overline{r_1}^{\,t} + B_2 A_{22}\overline{r_2}^{\,t}$$

We defined previously that $A_{11} = A_{12} = 1$. Substitution of the values of the characteristic roots into the expressions for A_{21} and A_{22} yields:

$$A_{21} = 0.2638$$

$$A_{22} = -1.2638$$

Substituting in the values for the characteristic roots, A_{ij} terms, and the particular solution obtained in (a) we obtain the general solution to the system of first-order difference equations.

$$y_{1t} \;=\; B_1(0.8791)^t + B_2(0.4209)^t + 17,857.14$$
$$y_{1t} \;=\; B_1 0.2638(0.8791)^t - B_2 1.2638(0.4209)^t + 7,571.43$$

Because the absolute value of both the characteristic roots is strictly less than one the system is stable and converges to the intertemporal equilibrium.

Chapter 12

Differential Equations

Objectives

The questions in this chapter should help the reader solve a variety of linear differential equations and appreciate their applications in economics. Readers who are able to answer all of the questions in the chapter should be able to:

1. Determine the degree and order of differential equations (Question 1).

2. Solve separable differential equations through integration (Question 2).

3. Solve exact differential equations and use an integrating factor to make a differential equation exact (Question 3).

4. Solve Bernoulli equations by transforming them into linear differential equations (Question 4).

5. Solve linear first-order differential equations using the general formula and determine whether the time path converges to the particular integral (Question 5).

6. Use phase diagrams to determine the stability of a dynamic system and obtain a qualitative solution to differential equations (Question 6).

7. Solve linear second-order differential equations, determine whether the time path converges to the particular integral and specify the definite solution given the initial conditions (Questions 7 and 8).

8. Solve a system of simultaneous linear differential equations (Question 9).

Review

Economic analysis is often concerned with how variables change over time. Differential equations can express the rates of change of variables in continuous time. A *solution* to a differential equation is the primitive function that is consistent with the original equation and which contains no derivative terms. A *definite* solution to a differential equation is obtained whenever the initial or boundary conditions or starting values of the variables are known.

Differential equations can be classified into several categories. The *order* of a differential equation is determined by the highest order of the derivative or differential in the differential equation.

The *degree* of a differential equation refers to the power to which the highest order derivative or differential is raised. The general form of an nth order differential equation is as follows:

$$f(x(t), \frac{dx}{dt}, \cdots + \frac{dx^n}{dt^n}) = w(t)$$

where if we set $w(t) = 0$ we obtain the *homogeneous* part of the differential equation.

A differential equation is *linear* if no variable, derivative, or differential is raised to a power other than one and if there is no derivative or differential term that is multiplied by a variable. If each coefficient that multiplies the derivative terms and dependent variable is a constant then the differential equation is linear with constant coefficients. The general form of a linear differential equation of order n with constant coefficients is as follows.

$$\frac{d^n x}{dt^n} + a_1 \frac{d^{n-1} x}{dt^{n-1}} \cdots + a_{n-1} \frac{dx}{dt} + a_n x = w(t)$$

where the differential equation is linear with constant coefficients and constant term if $w(t) = c$ where c is a constant.

A differential equation is *separable* if the independent and dependent variables can be expressed as separate functions, i.e.,

$$\frac{dx}{dt} = f(x)g(t)$$

Separable differential equations can be solved by separating the variables and then integrating both sides of the equation.

If the solution or primitive function is a function of more than one variable then we have a *partial* differential equation. A partial differential equation with two variables (x, y) is *exact* if the total differential of the primitive function $F(x, y)$ is zero and $\frac{\partial^2 F(x,y)}{\partial x \partial y} = \frac{\partial^2 F(x,y)}{\partial y \partial x}$ where the equality of the cross second-order partial derivatives is not assured. An exact differential equation may be nonlinear but will always be of order one and degree one. The method of solution is to partially integrate one of the partial derivatives while treating the other variable as a constant as in repeated integration. The function thus obtained is then differentiated with respect to the variable that was previously held constant and equated to the partial derivative of the variable in the actual partial differential equation. The next step is to obtain the integral of the unknown derivative function. Substituting this expression into the original integral we obtain the primitive function or solution to the differential equation.

Some partial differential equations that are not exact can be made exact by multiplying each term by a factor of integration. For the partial differential equation defined as follows:

$$\frac{\partial F}{\partial y} dy + \frac{\partial F}{\partial x} dx = dF(x, y)$$

where $\frac{\partial^2 F(x,y)}{\partial x \partial y} \neq \frac{\partial^2 F(x,y)}{\partial y \partial x}$. The integrating factor, if it exists, can be determined as follows:

<u>Rule One:</u>

Provided that

$$(\frac{\partial F}{\partial y})^{-1} [\frac{\partial^2 F(x,y)}{\partial x \partial y} - \frac{\partial^2 F(x,y)}{\partial y \partial x}] = a(x)$$

The integrating factor is $\exp^{\int a(x)dx}$.

<u>Rule Two:</u>

Provided that

$$(\frac{\partial F}{\partial x})^{-1}[\frac{\partial^2 F(x,y)}{\partial y \partial x} - \frac{\partial^2 F(x,y)}{\partial x \partial y}] = b(y)$$

The integrating factor is $\exp^{\int b(y)dy}$.

The rules for finding the integrating factor can be applied to the following partial differential equation which is not exact.

$$10xydx + (3x^2 + 4y)dy = 0$$

In this case we can define

$$\frac{\partial^2 F(x,y)}{\partial x \partial y} = 10x$$

$$\frac{\partial^2 F(x,y)}{\partial y \partial x} = 6x$$

Applying rule 1 we obtain

$$\frac{1}{3x^2 + 4y}(4x)$$

which is not a function of x alone and thus cannot be used to obtain the integrating factor.

Applying rule 2 we obtain

$$\frac{1}{10xy}(-4x) = -\frac{4}{10y}$$

which is a function y alone. The integrating factor is therefore $e^{\int \frac{-4}{10y}dy}$. Performing the integration and noting that $\ln x^a = a \ln x$ and $e^{\ln x} = x$ we obtain the following value for the integrating factor $\frac{1}{y^{\frac{4}{10}}}$. Multiplying every term in the partial differential equation by the integrating factor makes it exact.

A *Bernoulli* equation is a differential equation which is nonlinear in the dependent variable but can be transformed into a linear differential equation. A Bernoulli equation has the general form given below.

$$\frac{dx}{dt} = -g(t)x + f(t)x^n$$

where $g(t)$ and $f(t)$ may be either functions of the independent variable or constants. The Bernoulli equation is transformed into a linear differential equation by defining a new variable $z = x^{1-n}$ and rewriting the equation in terms of z as follows:

$$\frac{dz}{dt} = (1-n)[-g(t)z + f(t)]$$

The general solution to a differential equation where x is the dependent variable and t is the independent variable consists of a *complementary function* (x_c) which is the solution to the homogeneous part of the differential equation and a *particular integral* (x_p) which is the intertemporal equilibrium of $x(t)$.

A general formula can be applied to solve linear first-order differential equations. For a linear first-order differential equation,

$$\frac{dx}{dt} + a_1(t)x = g(t)$$

The general solution is as follows:

$$x(t) = e^{-\int a_1(t)dt}\left(A + \int g(t)e^{\int a_1(t)dt}dt\right)$$

where $a_1(t)$ and $g(t)$ may be either functions of the independent variable t or constants and A is an undetermined constant.

The complementary function is defined by $x_c = Ae^{-\int a_1(t)dt}$ and the particular integral by $x_p = e^{-\int a_1(t)dt}\int g(t)e^{\int a_1(t)dt}dt$. In the case where $a_1(t)$ and $g(t)$ are constants defined, respectively, as a_1 and c the complementary function is $x_c = Ae^{-a_1t}$ and the particular integral is $x_p = \frac{c}{a_1}$. In this case, if the initial condition defined by $x(0)$ is known, the definite solution is defined by $x_p = \frac{c}{a_1}$ and $x_c = (x(0) - \frac{c}{a_1})e^{-a_1t}$.

A general formula may also be applied to solve linear second-order differential equations with constant coefficients and a constant term. For the following differential equation,

$$\frac{d^2x}{dt^2} + a_1\frac{dx}{dt} + a_2x(t) = c$$

The particular integral is defined as follows:

$$
\begin{aligned}
x_p &= \frac{c}{a_2} \text{ given } a_2 \neq 0 \\
x_p &= \frac{c}{a_1}t \text{ given } a_1 \neq 0, a_2 = 0 \\
x_p &= \frac{c}{2}t^2 \text{ given } a_1 = a_2 = 0
\end{aligned}
$$

The complementary function has the following form,

$$
\begin{aligned}
x_c &= A_1e^{r_1t} + A_2e^{r_2t} \text{ if } r_1 \neq r_2 \\
x_c &= A_1e^{r_1t} + A_2te^{r_1t} \text{ if } r_1 = r_2
\end{aligned}
$$

where A_1 and A_2 are undetermined constants and r_1 and r_2 are the *characteristic roots* to the characteristic equation defined as the following quadratic equation.

$$r^2 + a_1r + a_2 = 0$$

The characteristic roots may be real and distinct, real and repeated or complex. If the characteristic roots are real the time path of the variable(s) converge to the particular solution if and only if both the roots are **negative**. In this case, the intertemporal equilibrium is said to be dynamically stable. In the case of two complex roots defined as the complex conjugate $(a \pm bi)$, a necessary condition for dynamic stability is that $a < 0$.

Convergence of an nth order linear differential equation with constant coefficients can be determined from the roots of the characteristic equation defined below.

$$r^n + a_1 r^{n-1} + \cdots + a_{n-1}r + a_n = 0$$

A necessary and sufficient condition for convergence to the particular integral is the *Routh-Hurwitz Theorem* which requires that the following n determinants are all positive.

$$\Delta_1 = a_1,$$

$$\Delta_2 = \begin{vmatrix} a_1 & a_3 \\ 1 & a_2 \end{vmatrix},$$

$$\Delta_3 = \begin{vmatrix} a_1 & a_3 & a_5 \\ 1 & a_2 & a_4 \\ 0 & a_1 & a_3 \end{vmatrix}, \cdots$$

$$\Delta_n = \begin{vmatrix} a_1 & a_3 & a_5 & \cdots & 0 \\ 1 & a_2 & a_4 & \cdots & 0 \\ 0 & a_1 & a_3 & \cdots & 0 \\ 0 & 1 & a_2 & \cdots & 0 \\ 0 & 0 & a_1 & \cdots & 0 \\ \cdots & \cdots & \cdots & \cdots & \cdots \\ \cdots & \cdots & \cdots & \cdots & a_n \end{vmatrix}$$

In applying the test for stability, all a_i terms where $i > n$ are treated as zero. Consequently, for a third order linear differential equation with constant coefficients the necessary and sufficient conditions for stability are

$$\Delta_1 = a_1 > 0,$$

$$\Delta_2 = \begin{vmatrix} a_1 & a_3 \\ 1 & a_2 \end{vmatrix} > 0,$$

$$\Delta_3 = \begin{vmatrix} a_1 & a_3 & 0 \\ 1 & a_2 & 0 \\ 0 & a_1 & a_3 \end{vmatrix} > 0$$

Differential equations can also be solved qualitatively. The method of solution is to use phase diagrams which graph the change in the dependent variable with respect to time against the dependent variable. To use a phase diagram the differential equation must be autonomous such that time—the independent variable—does not appear in the right hand side of the differential equation when written in the following way: $\frac{dx}{dt} = f(x)$.

Further Reading

This chapter provides an introduction to linear differential equations and should enable the reader to solve a number of economic problems. There is, however, a vast literature on linear and nonlinear differential equations. A good introduction to differential equations in economics is provided by Holden and Pearson [23] and Chiang [13] (chapters 14, 15 and 18). A highly recommended introduction to the topic is provided by Sydsæter and Hammond [32] (chapter 21). Dowling [17] (chapters 18 and 20) also provides numerous worked examples of first and second-order linear differential equations.

The interested reader may consult Brock and Malliaris [11], Baumol [3], Gandolfo [18], and Sydsæter [31] for further problems and applications of differential equations in economics at a more advanced level. Brock and Malliaris [11] address the issue of stability in chapters three and four while Gandolfo [18] in Appendix II and III provides a good discussion on the stability of linear and nonlinear systems and reviews the qualitative solution of nonlinear differential equations.

Chapter 12 - Questions

Degree and Order of Differential Equations

Question 1

Determine the order and degree of the following differential equations:

(i)

$$\frac{d^2y}{dt^2} + 4yt\frac{dy}{dt} = (\frac{d^2y}{dt^2})^2 .$$

(ii)

$$(\frac{d^2y}{dt^2})^3 - A(\frac{dy}{dt})^2 + yt = 0$$

(iii)

$$(\frac{d^2y}{dx^2})^4 + (\frac{d^5y}{dx^5})^2 - 10y = 0$$

(iv)

$$\frac{d^3y}{dw^3} + w^2y(\frac{d^2y}{dw^2})^4 - 20y^4 = 0$$

Separable Differential Equations

Question 2

Solve the following differential equations.

(i)

$$(3+x)dy = y^2dx$$

(ii)

$$\frac{dy}{dx}(xy + 2y) = (5x - xy^2)$$

(iii)

$$\frac{dy}{dt} = ty^2$$

(iv)

$$2t\frac{dy}{dt} = 4y$$

Exact Differential Equations

Question 3

Solve the following differential equations.

(i)

$$3x(xy - 2)dx + (x^3 + 2y)dy = 0$$

(ii)

$$(20x + 8y + 4)dx + (8x + 4y - 9)dy = 0$$

(iii)

$$(10y + 2x^2)dy + 2xydx = 0$$

Bernoulli Equations

Question 4

Solve the following differential equations.

(i)

$$\frac{dy}{dw} + 2y = wy^3$$

(ii)

$$\frac{dk}{dt} = sk^a - nk$$

General Formula

Question 5

The change in the capital stock in an economy over time is represented by the following differential equation.

$$\frac{dK}{dt} = abK(t) + x(t)$$

where $K(t)$ is the capital stock and $x(t)$ is direct foreign investment at time t and a and b are constants.

The time path of direct foreign investment is defined as:

$$x(t) = x(0)e^{\alpha t}$$

where $x(0)$ is direct foreign investment in the first time period, α is a constant and $ab - \alpha \neq 0$.

(a) Derive the time path of the capital stock in the economy.

(b) Under what conditions will the capital stock converge to an intertemporal equlibrium such that $K > 0$?

Phase Diagrams

Question 6

The change in capital (k) and output (q) over time in an economy are defined by the following differential equations.

(a) $\frac{dk}{dt} = q^{\frac{1}{2}} - k$

(b) $\frac{dq}{dt} = k - \frac{q}{5}$

(i) Find the values of k and q at the steady state.

(ii) Draw a phase diagram of $\frac{dk}{dt}$.

(iii) Draw a phase diagram of $\frac{dq}{dt}$.

(iv) Draw a diagram in (q, k) space where $\frac{dk}{dt} = 0$.

(v) Is the steady state found in (i) stable?

Linear Second-Order Differential Equations

Question 7

Suppose that savings (S) and investment (I) in an economy are such that investment is a function of both the change in income and whether income is rising and falling at an increasing or decreasing rate. The savings and investment functions at time t may be defined by the following equations where s is the marginal propensity to save out of income, c is autonomous savings, Y is national income, and i_1 and i_2 are constants.

$$\begin{aligned} S(t) &= c + sY(t) \\ I(t) &= i_1\frac{dY}{dt} + i_2\frac{d^2Y}{dt^2} \end{aligned}$$

(a) Assuming that savings equal investment, what is the intertemporal equilibrium for national income in the economy?

(b) Under what conditions would national income converge to the equilibrium? Do these conditions make economic sense?

(c) Solve for the time path of national income if $s = 0.025$, $i_1 = 0.2$, $i_2 = -0.5$ and given the initial conditions that national income at $t = 0$ is 10 and $\frac{dy}{dt} = 0.5$.

Question 8

Suppose the demand for Brent crude oil is determined by both a speculative component and a commercial component. The excess speculative demand for oil (ED_s) and excess commercial demand for oil (ED_c) at time t are represented by the following equations.

$$
\begin{aligned}
ED_s(t) &= d_1[EP(t) - P(t)] \\
ED_c(t) &= b_1 + b_2 P(t)
\end{aligned}
$$

where d_1, b_1, and b_2 are constants, $P(t)$ is the price of Brent crude oil, and $EP(t)$ is the expected future price of Brent crude oil formed by speculators.

Further suppose that $EP(t)$ is defined by the following equation.

$$
EP(t) = c_1 + c_2 P(t) + c_3 \frac{dP}{dt} + c_4 \frac{d^2 P}{dt^2}
$$

where c_1, c_2, c_3 and c_4 are constants.

(a) Solve for the intertemporal equilibrium of the price of Brent crude oil.

(b) Determine the conditions under which the time path is stable.

(c) Obtain a definite solution for the time path given that the price of Brent crude oil is 18 and $\frac{dP}{dt} = -2$ at $t = 0$ and that $\frac{c_3}{c_4} = 2$ and $\frac{d_1(c_2-1)+b_2}{d_1 c_4} = 1$.

Simultaneous System of Equations

Question 9

A dynamic input-output model of an industry that produces two goods (x_1) and (x_2) is presented below. x_1 is used an input in producing x_2 and x_2 is used as an input in producing x_1. In addition, a stock of x_1 is required as capital in producing x_2 and a stock of x_2 is required as capital to produce x_1. Investment or the change in the capital stock is linked to the change in output in the other good.

The outputs of x_1 and x_2 satisfy the following equations:

$$
\begin{aligned}
x_1 &= a_{11}x_1 + a_{12}x_2 + b_{11}\frac{dx_1}{dt} + b_{12}\frac{dx_2}{dt} + D_1 \\
x_2 &= a_{21}x_1 + a_{22}x_2 + b_{21}\frac{dx_1}{dt} + b_{22}\frac{dx_2}{dt} + D_1
\end{aligned}
$$

where a_{ij} is the input of good i currently used in the production of one unit of good j, b_{ij} is the stock of good i that is required to produce one unit of good j, and D_i is the final demand for good i.

(a) Derive the intertemporal equilibrium output for the two goods if $a_{11} = 0.2, a_{12} = 0.1, a_{21} = 0.4, a_{22} = 0.3, b_{11} = b_{22} = 0, b_{12} = 2, b_{21} = 1, D_1 = 10$, and $D_2 = 19$.

(b) Find the time paths of output of the two goods and determine if the dynamic system is stable.

Chapter 12 - Solutions

Degree and Order of Differential Equations

Question 1

(i)

The highest order derivative is two and the power to which the highest order derivative is raised is two. Thus the Order = 2 and Degree = 2.

(ii)

The highest order derivative is two and the power to which the highest order derivative is raised is three. Thus the Order = 2 and Degree = 3.

(iii)

The highest order derivative is five and the power to which the highest order derivative is raised is two. Thus the Order = 5 and Degree = 2.

(iv)

The highest order derivative is three and the power to which the highest order derivative is raised is one. Thus the Order = 3 and Degree = 1.

Separable Differential Equations

Question 2

(i)

This equation is not linear, not exact, but is separable in variables. The method of solution is to separate the variables such that one variable is on the LHS of the equation and the other is on the RHS of the equation and then integrate both sides.

$$(3 + x)dy = y^2 dx$$

$$\Rightarrow \frac{1}{y^2}dy = \frac{1}{3 + x}dx$$

Integrating both sides using the power rule and logarithmic rule of integration yields:

$$-y^{-1} + K_1 = \ln(3 + x) + K_2$$

$$\Rightarrow y^{-1} = -\ln(3 + x) + K$$

where $K = K_2 - K_1$ is a constant of integration. The differential equation is solved as it contains no derivative term.

(ii)

This equation is not linear, not exact, but can be shown to be separable in variables. The method of solution is to separate one variable to the LHS of the equation and the other variable to the RHS and then solve by integrating both sides of the equation.

$$dy(xy + 2y) = dx(5x - xy^2)$$

$$\Rightarrow dy \cdot xy + dy \cdot 2y = dx \cdot 5x - dx \cdot xy^2$$

$$\Rightarrow dy \cdot y(x + 2) = dx \cdot x(5 - y^2)$$

$$\Rightarrow \frac{y}{5 - y^2} \cdot dy = \frac{x}{x + 2} \cdot dx$$

Integrating both sides we obtain:

$$\int \frac{y}{5 - y^2} dy = \int \frac{x}{x + 2} dx$$

First let us integrate the LHS by defining $u = 5 - y^2$ and $\frac{du}{dy} = -2y$ to obtain

$$\int \frac{y}{5 - y^2} dy = \int \frac{-1}{2u} \frac{du}{dy} dy$$

Using the multiplication by a constant and the substitution rule of integration,

$$\int \frac{-1}{2u} \frac{du}{dy} dy = -\frac{1}{2} \int \frac{1}{u} du$$

Using the logarithmic rule of integration,

$$-\frac{1}{2} \int \frac{1}{u} du = -\frac{1}{2} \ln(u) + K_1$$

Integrating the RHS,

$$\int \frac{x}{x + 2} dx = \int (\frac{x + 2}{x + 2} - \frac{2}{x + 2}) dx$$

Using the logarithmic rule of integration yields

$$\int (\frac{x + 2}{x + 2} - \frac{2}{x + 2}) dx = x - 2 \ln(x + 2) + K_2$$

Equating the integral of the LHS and RHS we obtain

$$-\frac{1}{2}\ln(5 - y^2) + K_1 = x - 2\ln(x + 2) + K_2$$

$$\Rightarrow -\frac{1}{2}\ln(5 - y^2) = x - 2\ln(x + 2) + K$$

where $K = K_2 - K_1$ is a constant of integration. Remembering that $\ln(x^a) = a\ln(x)$.

$$\Rightarrow \ln(5 - y^2)^{-\frac{1}{2}} = x + \ln(x + 2)^{-2} + K$$

Taking the antilogs of both sides we obtain the solution:

$$(5 - y^2)^{-\frac{1}{2}} = e^x e^K (x + 2)^{-2}$$

$$\Rightarrow \frac{(x + 2)^2}{\sqrt{5 - y^2}} = e^{x+K}$$

(iii)

This equation is not linear, not exact, but is separable in variables.

$$\frac{dy}{dt} = ty^2$$

$$\Rightarrow \frac{dy}{y^2} = dt \cdot t$$

Integrating both sides using the power rule of integration yields:

$$\int \frac{1}{y^2} dy = \int t\, dt$$

$$\Rightarrow -y^{-1} + K_1 = \frac{t^2}{2} + K_2$$

$$\Rightarrow -y^{-1} = \frac{t^2}{2} + K$$

where $K = K_2 - K_1$ is a constant of integration.

Thus:

$$y^{-1} = -\frac{t^2}{2} - K$$

$$y^{-1} = \frac{-t^2 - 2K}{2}$$

$$y = \frac{2}{-t^2 - 2K}$$

(iv)

This equation is linear and separable in variables. The method of solution is to separate the variables such that one variable is on the LHS of the equation and the other is on the RHS of the equation.

$$2t\frac{dy}{dt} = 4y$$

$$\Rightarrow 2t \cdot dy = 4y \cdot dt$$

$$\Rightarrow \frac{dy}{4y} = \frac{dt}{2t}$$

Integrating both sides we obtain:

$$\int \frac{dy}{4y} = \int \frac{dt}{2t}$$

Using the multiplication by a constant rule of integration yields:

$$\frac{1}{4}\int \frac{1}{y}dy = \frac{1}{2}\int \frac{1}{t}dt$$

Using the logarithmic rule of integration yields

$$\frac{1}{4}\ln(y) + K_1 = \frac{1}{2}\ln(t) + K_2$$

$$\Rightarrow \ln y = 2 \cdot \ln t + K$$

where $K = 4(K_2 - K_1)$ is a constant of integration.

Taking antilogs and remembering that $\ln(x^a) = a\ln(x)$ we obtain

$$y(t) = t^2 \cdot e^K$$

We could also have solved the differential equation by noting that it is a linear homogeneous differential equation which can be rewritten as follows:

$$\frac{dy}{dt} - \frac{2y}{t} = 0$$

Using the general formula for a first-order linear differential equation we obtain:

$$y(t) = e^{-\int \frac{-2}{t}dt} \cdot A$$

$$\Rightarrow y(t) = e^{2\int \frac{1}{t}dt} \cdot A$$

$$\Rightarrow y(t) = e^{2\ln t} \cdot A$$

$$\Rightarrow y(t) = e^{\ln t^2} \cdot A$$

Remembering that $e^{\ln(x)} = x$ we obtain:

$$y(t) = t^2 \cdot A$$

where A is an arbitrary constant.

Exact Differential Equations

Question 3

(i)

This is an exact differential equation because $\frac{\partial^2 F(x,y)}{\partial x \partial y} = \frac{\partial^2 F(x,y)}{\partial y \partial x}$ where $F(x,y)$ is the primitive function, i.e.,

$$\frac{d}{dy}[3x(xy-2)] = \frac{d}{dx}[x^3 + 2y] = 3x^2$$

First, integrate with respect to the variable x treating y as a constant.

$$F(y,x) = \int [3x(xy-2)]dx + \phi(y)$$

where $\phi(y)$ is an unknown function of the variable y. Evaluating the integral yields:

$$F(y,x) = x^3 y - 3x^2 + \phi(y)$$

Taking the partial derivative of the $F(x,y)$ with respect y, the variable previously held constant, we obtain:

$$\frac{\partial F}{\partial y} = x^3 + \phi'(y)$$

This expression may be compared to the partial derivative of the actual primitive function with respect to y which is given in the original differential equation.

$$\frac{\partial F}{\partial y} = x^3 + 2y$$

Both terms must be equal such that:

$$x^3 + \phi'(y) = x^3 + 2y$$

Solving for the unknown $\phi'(y)$ function we obtain:

$$\phi'(y) = 2y$$

Integrating $\phi'(y)$ we can obtain the unknown function $\phi(y)$.

$$\int \phi'(y)dy = \phi(y) = y^2 + K$$

Substituting back $\phi(y)$ into the expression for $F(y,x)$ found by first integrating with respect to x but holding y constant we obtain an expression for the primitive function $F(x,y)$.

$$F(y,x) = x^3 y - 3x^2 + y^2 + K$$

where K is an arbitrary constant of integration.

(ii)

This is an exact differential equation because $\frac{\partial^2 F(x,y)}{\partial x \partial y} = \frac{\partial^2 F(x,y)}{\partial y \partial x}$ where $F(x,y)$ is the primitive function, i.e.,

$$\frac{d}{dy}[20x + 8y + 4)] = \frac{d}{dx}[8x + 4y - 9] = 8$$

Integrating with respect to the variable x while treating y as a constant yields:

$$F(x,y) = \int (20x + 8y + 4)dx + \phi(y)$$

$$\Rightarrow F(x,y) = 10x^2 + 8xy + 4x + \phi(y)$$

Partially differentiating the function $F(x,y)$ with respect to y we obtain:

$$\frac{\partial F(x,y)}{\partial y} = 8x + \phi'(y)$$

From the differential equation the partial derivative of $F(x, y)$ with respect to y is

$$\frac{\partial F(x, y)}{\partial y} = 8x + 4y - 9$$

Setting the two expressions equal to each other and solving out for the unknown expression $\phi'(y)$ yields:

$$8x + \phi'(y) = 8x + 4y - 9$$

$$\Rightarrow \phi'(y) = 4y - 9$$

Integrating $\phi'(y)$ we obtain the following expression:

$$\int \phi'(y) dy = \phi(y) = 2y^2 - 9y + K$$

Substituting $\phi(y)$ into the expression for F(x,y) yields the primitive function or solution to the differential equation.

$$F(x, y) = 10x^2 + 8xy + 4x + 2y^2 - 9y + K$$

where K is a constant of integration.

(iii)

This equation is not linear, not separable in variables, and is not exact. The equation may be made exact by the use of an *integrating factor*. First, we may observe that:

$$\frac{d}{dx}(10y + 2x^2) = 4x$$

$$\frac{d}{dy}(2xy) = 2x$$

If we multiply the differential equation by the common factor y then the differential equation will be exact. Performing the multiplication we obtain:

$$(10y^2 + 2x^2 y) \cdot dy + 2xy^2 \cdot dx = 0$$

The transformed differential equation is exact because $\frac{\partial^2 F(x,y)}{\partial x \partial y} = \frac{\partial^2 F(x,y)}{\partial y \partial x}$ where $F(x, y)$ is the primitive function, i.e.,

$$\frac{d}{dx}(10y^2 + 2x^2 y) = 4xy$$

$$\frac{d}{dy}(2xy^2) = 4xy$$

To solve the exact differential equation we must first integrate the equation with respect to one of the variables treating the other as a constant. Integrating with respect to y we obtain

$$F(x,y) = \int (10y^2 + 2x^2y)dy + \phi(x)$$

$$\Rightarrow F(x,y) = \frac{10}{3}y^3 + x^2y^2 + \phi(x)$$

The partial derivative of $F(x,y)$ with respect to the variable x is

$$\frac{\partial F(x,y)}{\partial x} = 2xy^2 + \phi'(x)$$

From the exact differential equation the partial derivative of $F(x,y)$ with respect to x is

$$\frac{\partial F(x,y)}{\partial x} = 2xy^2$$

Equating the two expressions and solving for $\phi'(x)$ yields,

$$2xy^2 + \phi'(x) = 2xy^2$$

$$\Rightarrow \phi'(x) = 0$$

Integrating $\phi'(x)$ we obtain

$$\int \phi'(x)dx = \phi(x) = K$$

where K is a constant of integration.

Substituting the expression for $\phi(x)$ into $F(x,y)$ we obtain the primitive function or solution to the differential equation.

$$F(x,y) = \frac{10}{3}y^3 + x^2y^2 + K$$

Bernoulli Equations

Question 4

(i)

This equation is neither linear, nor exact, nor separable in the variables. It is possible, however, to transform the equation into a linear first-order differential equation and it is thus a Bernoulli equation.

The method of solution is to first transform the equation so that it is linear and then solve the equation using the general formula for a linear first-order differential equation. To transform the equation we should divide through by y^n where y is the dependent variable of the primitive function and is raised to the power of n in the Bernoulli equation.

$$y^{-3} \cdot \frac{dy}{dw} + 2y^{-2} = w$$

If we define a new variable such that $z = y^{1-n} = y^{-2}$ we can rewrite the equation as a linear first-order differential equation.

$$-\frac{1}{2} \cdot \frac{dz}{dw} + z = w$$

It can be verified that the equations in terms of y and w and z and w are identical by differentiating z with respect to w using the chain rule of differentiation and substituting the expression for $\frac{dz}{dw}$ into the modified equation.

$$\frac{dz}{dw} = \frac{dz}{dy} \cdot \frac{dy}{dw} = -2y^{-3} \frac{dy}{dw}$$

If we multiply the modified differential equation by -2 we obtain a linear first-order differential equation that may be solved using the general formula.

$$\frac{dz}{dw} - 4z = -2w$$

Applying the general formula we obtain

$$z(w) = e^{\int 4dw} \cdot \left(A + \int -2we^{-\int 4dw} dw \right)$$

$$\Rightarrow z(w) = e^{4w} \cdot \left(A - 2 \int we^{-4w} dw \right)$$

To integrate the expression $\int we^{-4w} dw$ we can employ the integration by parts rule, i.e.,

$$\int u(w)v'(w)dw = u(w)v(w) - \int v(w)u'(w)dw$$

Defining $u(w)$ and $v(w)$ appropriately we obtain

$$u(w) = w$$

$$\Rightarrow u'(w) = 1$$

$$v'(w) = e^{-4w}$$

$$\Rightarrow \int v'(w)dw = v(w) = -\frac{1}{4}e^{-4w}$$

Integrating the expression we obtain

$$\int we^{-4w}dw = -\frac{1}{4}e^{-4w}w - \int -\frac{1}{4}e^{-4w}dw$$

$$\Rightarrow \int we^{-4w}dw = -\frac{1}{4}e^{-4w}w - \frac{1}{16}e^{-4w}$$

Substituting the expression for $\int we^{-4w}$ into the expression obtained from the general formula yields:

$$z(w) = e^{4w} \cdot \left(A + \frac{w}{2}e^{-4w} + \frac{1}{8}e^{-4w}\right)$$

$$\Rightarrow z(w) = Ae^{4w} + \frac{w}{2} + \frac{1}{8}$$

To obtain the solution in terms of the original variables y and w we must substitute $z(w) = y(w)^{-2}$. This yields

$$y^{-2}(w) = Ae^{4w} + \frac{w}{2} + \frac{1}{8}$$

$$\Rightarrow y(w) = \left(Ae^{4w} + \frac{w}{2} + \frac{1}{8}\right)^{-\frac{1}{2}}$$

(ii)

This equation is the fundamental equation from neo-classical growth theory where the economy is assumed to have just two factors of production, capital and labor. For a given savings rate s and growth rate in the population of n one can obtain the above differential equation as a representation of the economy where k is the per capita capital stock.

The equation is neither linear, nor exact, nor separable in the variables. It is possible, however, to transform it into a linear first-order differential equation. It is, therefore, a Bernoulli equation.

The method of solution is to first transform the equation so that it is linear and then solve the equation using the general formula for a linear first-order differential equation. To transform the equation we should divide through by k^a where k is the dependent variable of the primitive function.

$$\frac{dk}{dt}k^{-a} + nk^{1-a} = s$$

If we define a new variable such that $x = k^{1-a}$ we can rewrite the equation as a linear first-order differential equation.

$$\frac{dx}{dt} + nx(1-a) = s(1-a)$$

It can be verified that the equations in terms of k and t and x and t are identical by differentiating x with respect to t using the chain rule of differentiation and substituting into the modified equation the expression for $\frac{dx}{dt}$.

$$\frac{dx}{dt} = \frac{dx}{dk} \cdot \frac{dk}{dt} = (1-a)k^{-a}\left(\frac{dk}{dt}\right)$$

Applying the general formula to the modified equation in terms of x and t we obtain

$$x(t) = e^{\int (a-1)ndt} \cdot \left(A + \int (1-a)se^{\int (1-a)ndt}dt\right)$$

$$\Rightarrow x(t) = e^{(a-1)nt} \cdot \left(A + \int (1-a)se^{(1-a)nt}dt\right)$$

$$\Rightarrow x(t) = e^{(a-1)nt} \cdot \left(A + (1-a)s\int e^{(1-a)nt}dt\right)$$

$$\Rightarrow x(t) = e^{(a-1)nt} \cdot \left(A + \frac{s}{n}e^{(1-a)nt}\right)$$

$$\Rightarrow x(t) = Ae^{(a-1)nt} + \frac{s}{n}$$

where A is an undetermined constant. Substituting back $x = k^{1-a}$ we obtain:

$$k(t) = \left(Ae^{(a-1)nt} + \frac{s}{n}\right)^{\frac{1}{1-a}}$$

The solution represents the time path of the per capita capital stock in the economy. Given that $a < 1$ as $t \Rightarrow \infty$ then $k(t)$ converges to $\left(\frac{s}{n}\right)^{\frac{1}{1-a}}$. Thus, the equilibrium level of k (which determines the equilibrium level of per capita income) is increasing in the savings rate and decreasing in the population growth rate.

General Formula

Question 5

(a) Substituting the expression for x_t into the differential equation for the capital stock we obtain the following differential equation.

$$\frac{dK}{dt} - abK(t) = x(0)e^{\alpha t}$$

This is a linear first-order differential equation with constant coefficients. The equation can be solved to obtain the primitive function, or the time path of the capital stock, by applying the general formula:

$$K(t) = e^{-\int -abdt}(A + \int x(0)e^{\alpha t}e^{\int -abdt}dt)$$

$$\Rightarrow K(t) = e^{abt}(A + x(0)\int e^{\alpha t}e^{-abt}dt)$$

$$\Rightarrow K(t) = e^{abt}(A + x(0)\int e^{(\alpha - ab)t}dt)$$

$$\Rightarrow K(t) = e^{abt}(A + \frac{x(0)}{\alpha - ab}e^{(\alpha - ab)t})$$

$$\Rightarrow K(t) = Ae^{abt} + \frac{x(0)}{\alpha - ab}e^{\alpha t}$$

(b) Stability of the time path capital requires that small deviations in the initial conditions diminish over time such that as $t \to \infty$ the time path converges to the particular integral. The necessary and sufficient conditions for stability are that $ab < 0$ and $\alpha < 0$.

Phase Diagrams

Question 6

(i) At the steady state of k it must be the case that:

$$\frac{dk}{dt} = 0$$

$$\Rightarrow q^{\frac{1}{2}} = k$$

At the steady state of q it must be the case that:

$$\frac{dq}{dt} = 0$$

$$\Rightarrow \frac{q}{5} = k$$

Combining the two results we obtain:

$$q^{\frac{1}{2}} = \frac{q}{5}$$

$$\Rightarrow q^* = 25$$

where q^* is the steady state of output. It follows that if $q^* = 25$ the steady state of the capital stock k^* is 5.

(ii) Phase diagrams give us a qualitative solution to a differential equation without actually solving for the time path of the variables provided the differential equations are autonomous. In economics we are often concerned whether there exist equilibrium states and whether the equilibria are stable. A phase diagram provides us with this information.

To draw the phase line of the differential equation on the phase diagram we note that if $\frac{dk}{dt} > 0$ then $k(t)$ must be increasing with time so we move from the left to the right in the diagram. If $\frac{dk}{dt} < 0$ then $k(t)$ must be decreasing with time so we move from the right to the left in the diagram. The arrows on the phase line indicate this direction. An *equilibrium* is found whenever $\frac{dk}{dt} = 0$ such that $k(t)$ is stationary. An equilibrium is locally *stable* if when we move a little to the right or the left of the equilibrium it converges to the equilibrium. It can be seen from the phase diagrams of both capital and output that for an equilibrium to be stable the phase line must have a **negative** slope at the equilibrium. This concurs with a quantitative solution to a differential equation that for convergence to the intertemporal equilibrium we require that all the real roots are negative.

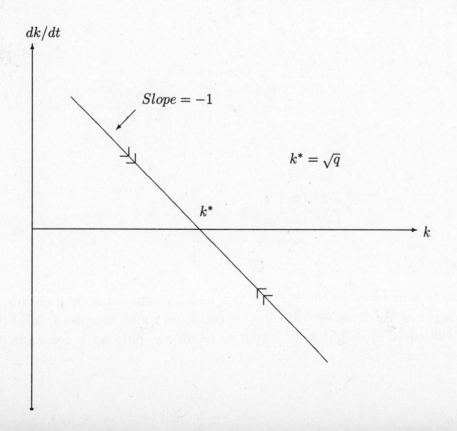

(iii)

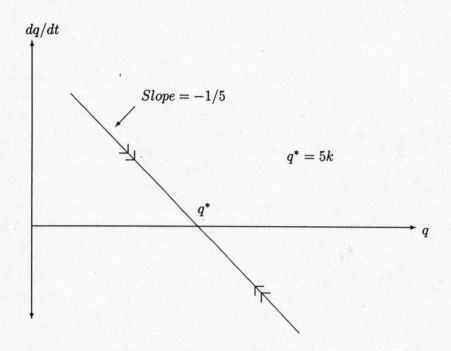

(iv) To find $\frac{dk}{dt} = 0$ in (k,q) space we note from (a) that if $\frac{dk}{dt} = 0$ then $k = q^{\frac{1}{2}}$. This function $k = k(q)$ is strictly concave and increasing in q given that

$$\frac{dk(q)}{dq} = \frac{1}{2}q^{-\frac{1}{2}} > 0$$

$$\frac{d^2k(q)}{dq^2} = -\frac{1}{4}q^{-\frac{3}{2}} < 0$$

The curve $\frac{dk}{dt} = 0$ is represented by the $k(q)$ function and is illustrated in figure below. We can examine the behavior of $\frac{dk}{dt}$ off the $\frac{dk}{dt} = 0$ curve by noting that if we increase k, **holding q constant**, then we will reduce $\frac{dk}{dt}$ such that $\frac{dk}{dt} < 0$. If we increase q, **holding k constant**, then we will increase $\frac{dk}{dt}$ such that $\frac{dk}{dt} > 0$.

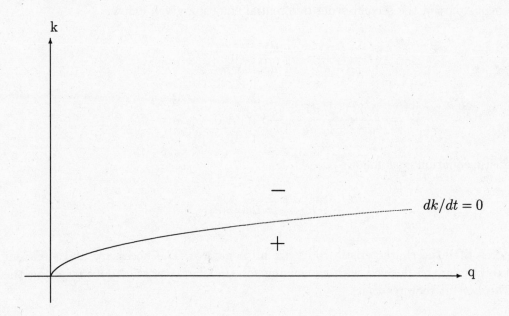

(v) From the phase diagrams given in (ii) and (iii), we can see that whether k is less than or greater than k^* it always converges to the steady state. Similarly, if q is less than or greater than q^* it also converges to the steady state. Thus, the steady state is stable.

Linear Second-Order Differential Equations

Question 7

(a)

To determine the intertemporal equilibrium or particular integral we must first set $S(t) = I(t)$, i.e.,

$$sY(t) - i_1 \frac{dY}{dt} - i_2 \frac{d^2Y}{dt^2} = -c$$

One possible equilibrium is that Y is constant such that $\frac{dY}{dt} = \frac{d^2Y}{dt^2} = 0$. Setting $Y(t) = Y_e$ we obtain the solution

$$sY_e = c$$

$$\Rightarrow Y_e = \frac{-c}{s}$$

(b)

To determine the conditions under which the level of income converges to the intertemporal equilibrium we must first obtain the characteristic equation. The characteristic equation is derived from the homogeneous part of the second-order differential equation given below.

$$sY - i_1 \frac{dY}{dt} - i_2 \frac{d^2Y}{d^2} = 0$$

$$\Rightarrow \frac{d^2Y}{dt^2} + \frac{i_1}{i_2}\frac{dY}{dt} - \frac{s}{i_2}Y(t) = 0$$

The characteristic equation is as follows

$$r^2 + \frac{i_1}{i_2}r - \frac{s}{i_2} = 0$$

Stability requires that the characteristic roots are all negative. The necessary and sufficient conditions for stability can be derived without solving for the characteristic roots by using the *Routh-Hurwitz Theorem*. It requires that

$$\Delta_1 = \frac{i_1}{i_2} > 0,$$

$$\Delta_2 = \begin{vmatrix} \frac{i_1}{i_2} & 0 \\ 1 & -\frac{s}{i_2} \end{vmatrix} > 0 \Rightarrow -\frac{si_1}{i_2^2} > 0$$

The marginal propensity to save (s) is positive and economic logic would suggest that $i_1 > 0$ such that investment in the economy is increasing with the rate of change in national income. The stability conditions to ensure convergence to an intertemporal equilibrium, however, imply that either s or i_1 must be negative.

(c)

Substituting the values for s, i_1, and i_2 into the characteristic equation we obtain:

$$r^2 - 0.4r + 0.05 = 0$$

It can be determined that there are two complex roots because $0.4^2 - 4 \cdot 0.05 = -0.04 < 0$ (see chapter 9). The complex roots can be solved as follows using the quadratic formula.

$$r_1, r_2 = \frac{0.4 \pm \sqrt{0.4^2 - 4 \cdot 0.05}}{2}$$

$$\Rightarrow r_1, r_2 = 0.2 \pm \frac{\sqrt{0.04}\sqrt{-1}}{2}$$

Defining $\sqrt{-1} = i$ we obtain the complex conjugate

$$r_1, r_2 = 0.2 \pm 0.1i$$

The general form of the complementary function (y_c) where the roots of the characteristic equation are a complex conjugate defined by $a \pm bi$ is

$$y_c = e^{at}(A_1 \cos bt + A_2 \sin bt)$$

where A_1 and A_2 are arbitrary and undetermined constants.

Performing the substitutions where $a = 0.2$ and $b = 0.1$ we obtain the complementary function.

$$y_c = e^{0.2t}(A_1 \cos 0.1t + A_2 \sin 0.1t)$$

where A_1 and A_2 are arbitrary and undetermined constants.

The time path of national income is the sum of the complementary function and the particular integral.

$$Y(t) = e^{0.2t}(A_1 \cos 0.1t + A_2 \sin 0.1t) - \frac{c}{s}$$

The definite solution can be found by determining the values for A_1 and A_2. The initial conditions are that $Y(0) = 10$ and $\frac{dY}{dt} = 0.5$ at $t = 0$. Substituting $t = 0$ into the general solution and $Y(0) = 10$ and noting that if $\theta = 0$ then $\cos \theta = 1$ and $\sin \theta = 0$ we obtain,

$$10 = A_1 - \frac{c}{s}$$

$$\Rightarrow A_1 = 10 + \frac{c}{s}$$

Differentiating the general solution with respect to t and then substituting in the value $\frac{dY}{dt} = 0.5$ at $t = 0$ we can solve for A_2. Using the product and chain rules of differentiation and noting that $\frac{d}{d\theta} \cos \theta = -\sin \theta$ and $\frac{d}{d\theta} \sin \theta = \cos \theta$ we obtain

$$\frac{dY}{dt} = 0.2e^{0.2t}(A_1 \cos 0.1t + A_2 \sin 0.1t) + e^{0.2t}[A_1(-0.1 \sin 0.1t) + A_2 0.1 \cos 0.1t]$$

Evaluating at $t = 0$ and noting that $\frac{dY}{dt} = 0.5$ yields

$$0.5 = 0.2A_1 + 0.1A_2$$

$$\Rightarrow A_2 = 5 - 2A_1$$

Substituting for the previously found value of A_1 we obtain

$$A_2 = -15 - \frac{2c}{s}$$

Thus the definite solution is

$$Y(t) = e^{0.2t}[(10 + \frac{c}{s}) \cos 0.1t + (-15 - \frac{2c}{s}) \sin 0.1t] - \frac{c}{s}$$

The time path for the economy, given the initial conditions and setting autonomous savings (c) equal to -1, is presented in the figure below. An infinite number of time paths exist depending upon the value we choose for c.

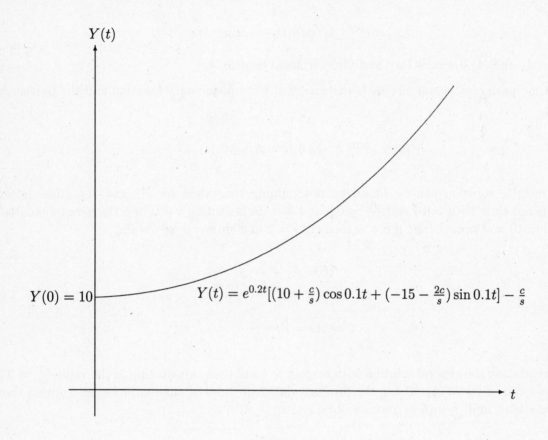

Question 8

(a)

At an equilibrium the sum of the excess demands is zero, i.e.,

$$ED_s + ED_c = 0$$

$$\Rightarrow d_1[EP(t) - P(t)] + b_1 + b_2P(t) = 0$$

$$\Rightarrow d_1(c_1 + c_2P(t) + c_3\frac{dP}{dt} + c_4\frac{d^2P}{dt^2}) - d_1P(t) + b_1 + b_2P(t) = 0$$

$$\Rightarrow d_1 c_4 \frac{d^2 P}{dt^2} + d_1 c_3 \frac{dP}{dt} + [d_1(c_2 - 1) + b_2]P(t) = -d_1 c_1 - b_1$$

$$\Rightarrow \frac{d^2 P}{dt^2} + \frac{c_3}{c_4}\frac{dP}{dt} + \frac{[d_1(c_2 - 1) + b_2]}{d_1 c_4}P(t) = -\frac{(d_1 c_1 + b_1)}{d_1 c_4}$$

One possible solution for the intertemporal equilibrium is that the price is constant such that $P(t) = P_e$ and $\frac{dP}{dt} = \frac{d^2 P}{dt^2} = 0$. Substituting this solution into the linear second-order differential equation we obtain

$$\frac{[d_1(c_2 - 1) + b_2]}{d_1 c_4}P_e = -\frac{(d_1 c_1 + b_1)}{d_1 c_4}$$

$$\Rightarrow P_e = -\frac{(d_1 c_1 + b_1)}{(d_1(c_2 - 1) + b_2)}$$

(b)

The stability condition may be obtained without solving for the characteristic roots by using the *Routh-Hurwitz Theorem*. First, we obtain the characteristic equation from the homogeneous part of the differential equation.

$$r^2 + \frac{c_3}{c_4}r + \frac{[d_1(c_2 - 1) + b_2]}{d_1 c_4} = 0$$

The necessary and sufficient conditions for stability are that the following determinants are all positive.

$\Delta_1 = \frac{c_3}{c_4} > 0,$

$$\Delta_2 = \begin{vmatrix} \frac{c_3}{c_4} & 0 \\ 1 & \frac{d_1(c_2-1)+b_2}{d_1 c_4} \end{vmatrix} > 0 \Rightarrow \frac{c_3[d_1(c_2-1)+b_2]}{d_1 c_4^2} > 0$$

(c)

Given that $\frac{c_3}{c_4} = 2$ and $\frac{d_1(c_2-1)+b_2}{d_1 c_4} = 1$ the characteristic equation is

$$r^2 + 2r + 1 = 0$$

Given that $2^2 - 4 \cdot 1 = 0$ there must be two repeated real roots. The roots may be obtained from the quadratic formula and are given below.

$$r_1 = r_2 = \frac{-2}{2} = -1$$

The general solution to the differential equation is the sum of the complementary function and particular integral. The particular integral was obtained in (a) and the general form of the complementary function is given below.

$$P_c = A_1 e^{r_1 t} + A_2 t e^{r_1 t}$$

where A_1 and A_2 are arbitrary and unknown constants.

Thus the general solution to the time path of the price of Brent crude oil is

$$P(t) = A_1 e^{-t} + A_2 t e^{-t} - \frac{(d_1 c_1 + b_1)}{(d_1(c_2 - 1) + b_2)}$$

Substituting in the value of $P(0) = 18$ and evaluating at $t = 0$ we obtain

$$18 = A_1 - \frac{(d_1 c_1 + b_1)}{(d_1(c_2 - 1) + b_2)}$$

$$\Rightarrow A_1 = 18 + \frac{(d_1 c_1 + b_1)}{(d_1(c_2 - 1) + b_2)}$$

Differentiating the general solution we obtain

$$\frac{dP}{dt} = -A_1 e^{-t} + A_2 e^{-t} - A_2 t e^{-t}$$

Substituting the other initial condition that at $t = 0$ $\frac{dP}{dt} = -2$ and evaluating at $t = 0$ we obtain

$$-2 = -A_1 + A_2$$

$$\Rightarrow A_2 = A_1 - 2$$

Substituting the previously found value of A_1 we obtain

$$A_2 = 16 + \frac{(d_1 c_1 + b_1)}{(d_1(c_2 - 1) + b_2)}$$

Thus the definite solution to the time path of the price of Brent crude oil is

$$P(t) = [18 + \frac{(d_1 c_1 + b_1)}{(d_1(c_2 - 1) + b_2)}]e^{-t} + [16 + \frac{(d_1 c_1 + b_1)}{(d_1(c_2 - 1) + b_2)}]t e^{-t} - \frac{(d_1 c_1 + b_1)}{(d_1(c_2 - 1) + b_2)}$$

Simultaneous System of Equations

Question 9

(a)

Substituting in the a_{ij}, b_{ij} and D_i values into the dynamic input-output model we obtain

$$x_1 = 0.2x_1 + 0.1x_2 + 2\frac{dx_2}{dt} + 10$$

$$x_2 = 0.4x_1 + 0.3x_2 + \frac{dx_1}{dt} + 19$$

Collecting terms yields,

$$0.8x_1 - 0.1x_2 - 2\frac{dx_2}{dt} = 10$$

$$-0.4x_1 + 0.7x_2 - \frac{dx_1}{dt} = 19$$

One possible solution for the intertemporal equilibrium is that both x_1 and x_2 are constant such that $x_1 = x_1^e$ and $x_2 = x_2^e$ and $\frac{dx_1}{dt} = \frac{dx_2}{dt} = 0$. Substituting this trial solution into the two equations we obtain

$$0.8x_1^e - 0.1x_2^e = 10$$
$$-0.4x_1^e + 0.7x_2^e = 19$$

From the first equation we obtain

$$-0.05x_2^e - 5 = -0.4x_1^e$$

Substituting into the second equation and solving we obtain $x_2^e = 32$. Substituting back into the first equation we obtain $x_1^e = 16.5$. This is consistent with the dynamic input-output model and is, therefore, the intertemporal equilibrium.

(b)

To solve for the time path of output for the goods we need to obtain the complementary function. From the solution of linear first-order differential equations we may adopt the following possible solution for the complementary functions where r is the characteristic root and the complementary functions are the solutions to the homogeneous part of the differential equations.

$$x_1(t) = A_1 e^{rt}$$

$$x_2(t) = A_2 e^{rt}$$

where A_1 and A_2 are unknown constants.

Substituting this trial solution into the homogeneous part of the two differential equations and noting that $\frac{d}{dt}e^{rt} = re^{rt}$ we obtain

$$e^{rt}[A_1 0.8x_1 - A_2 0.1x_2 - A_2 2r] = 0$$
$$e^{rt}[A_1 - 0.4x_1 + A_2 0.7x_2 - A_1 r] = 0$$

The trial solutions will only be a solution to the system of linear differential equations if the above conditions are satisfied. This requires that:

$$A_1 0.8x_1 - A_2 0.1x_2 - A_2 2r = 0$$
$$-A_1 0.4x_1 + A_2 0.7x_2 - A_1 r = 0$$

To obtain the characteristic equation we can exclude the trivial case where $A_1 = A_2 = 0$. Next, we can define the conditions in matrix form.

$$R = \begin{bmatrix} 0.8 & -0.1 - 2r \\ -0.4 - r & 0.7 \end{bmatrix},$$

$$A = \begin{bmatrix} A_1 \\ A_2 \end{bmatrix}$$

In matrix form the conditions may be written as follows:

$$RA = 0$$

The matrix R must be singular if there is to be a non-trivial solution to the set of conditions. From our knowledge of matrix algebra (see chapter 2) this requires that the determinant of the matrix R equals zero, i.e.,

$$R = \begin{vmatrix} 0.8 & -0.1 - 2r \\ -0.4 - r & 0.7 \end{vmatrix} = 0$$

Expanding the determinant of the matrix R yields the characteristic equation.

$$-2r^2 - 0.9r + 0.52 = 0$$

$$\Rightarrow r^2 + 0.45r - 0.26 = 0$$

Because $0.45^2 - 4 \cdot (-0.26) > 0$ there are two real distinct roots that solve the characteristic equation. Solving for the roots using the quadratic formula we obtain

$$r_1, r_2 = \frac{-0.225 \pm \sqrt{0.45^2 - 4 \cdot (-0.26)}}{2}$$

$$\Rightarrow r_1 = 0.3323374, \quad r_2 = 0.7823374$$

It can be shown that both r_1 and r_2 satisfy the original set of conditions imposed by the trial solution. Consequently, there are two sets of solutions. In the case where $r = r_1$ we may denote the solution by A_{11} and A_{21}.

$$A_{11}0.8x_1 - A_{21}0.1x_2 - A_{21}2r = 0$$
$$-A_{11}0.4x_1 + A_{21}0.7x_2 - A_{11}r = 0$$

With no loss in generality we can arbitrarily fix $A_{11} = 1$ and solve for A_{21} in terms of r_1 and the coefficients and obtain the following.

$$A_{21} = \frac{0.8}{(0.1 + r_1)} = \frac{r_1 + 0.4}{0.7}$$

Likewise if we substitute r_2 into the set of conditions and define the solution by A_{12} and A_{22} and set $A_{12} = 1$ we obtain an expression for A_{22}.

$$A_{22} = \frac{0.8}{(0.1 + r_2)} = \frac{r_2 + 0.4}{0.7}$$

A general solution for the outputs of the two goods must include the two solutions obtained from the two characteristic roots. If we define B_1 and B_2 as two unknown constants a general solution to the homogeneous part of the dynamic system is as follows:

$$x_1 = B_1A_{11}e^{r_1t} + B_2A_{12}e^{r_2t}$$
$$x_2 = B_1A_{21}e^{r_1t} + B_2A_{22}e^{r_2t}$$

We defined previously that $A_{11} = A_{12} = 1$. Substitution of the values of the characteristic roots into the expressions for A_{21} and A_{22} yields:

$$A_{21} = 1.046196$$

$$A_{22} = -0.5461963$$

Substituting in the values for the characteristic roots, A_{ij} terms, and the particular integral obtained in (a) we obtain the general solution to the system of linear first-order differential equations.

$$x_1(t) = B_1e^{0.3323t} + B_2e^{-0.7823t} + 16.5$$
$$x_2(t) = B_11.0462e^{0.3323t} - B_20.5462e^{-0.7823t} + 32$$

Because one of the characteristic roots is **not** negative the simultaneous system is not stable and will not converge to the particular integral.

Glossary

adjoint matrix: The adjoint matrix is the transpose of the matrix of cofactors.

argument: The independent variable in a function $y = f(x)$ is x, the argument of the function.

Arrow-Enthoven conditions: The conditions in a nonlinear programming problem that will ensure that if the constraint qualification is satisfied the Kuhn-Tucker conditions are both necessary and sufficient for a maximum.

artificial variables: These are variables created when solving linear programming problems using the simplex method. The variables are a device to ensure an initial basis and are assigned a very high negative (positive) value for maximization (minimization) problems.

augmented coefficient matrix: The matrix formed from the matrix of technical coefficients ($\mathbf{A}$) and the vector of constants ($\mathbf{b}$) for a system of simultaneous equations. Thus, if the system is defined as $\mathbf{Ax} = \mathbf{b}$ then the augmented coefficient matrix is $[\mathbf{A}|\mathbf{b}]$.

autonomous differential equation: A differential equation where time—the independent variable—does not appear in the right hand side of the differential equation when it is written in the following form $\frac{dy}{dt} = f(y)$.

Bernoulli equation: A Bernoulli equation is a differential equation nonlinear in the dependent variable and has the form:

$$\frac{dx}{dt} = -g(t)x + f(t)x^n$$

Bernoulli equations can be transformed into linear differential equations.

binding constraints: If a constraint effectively imposes a restriction on the feasible solution for a given problem then we refer to that constraint as a binding constraint.

border: The rows and columns of a bordered Hessian matrix in a constrained optimization problem which are the second-order partial derivatives of the Lagrangean multipliers and the original choice variables.

bordered Hessian: The matrix formed from the second-order partial derivatives from the Lagrangean function.

bounded set: A set is bounded if the whole set can be contained within a sufficiently large enough circle.

characteristic equation: Is formed from equations of the form $\mathbf{Ax} = \lambda\mathbf{x}$ and is defined as $\det(\mathbf{A} - \lambda\mathbf{I}) = 0$ where $\mathbf{x}$ is a non-zero vector. The values for λ that satisfy the characteristic equation are called characteristic roots or eigenvalues.

characteristic roots: Also called eigenvalues are the roots to the characteristic equation.

choice variables: Variables that individuals choose when solving optimization problems.

closed set: A set which does contains all its boundary points. The set $[0, 1]$ is a closed set as it contains all values between 0 and 1 and the boundary points, 0 and 1.

cofactors: The cofactor of the element a_{ij} of a matrix $\mathbf{A}$ is defined as c_{ij} and is the product of $(-1)^{i+j}$ and the minor of a_{ij}.

compact set: A set that is both bounded and closed is sometimes known as a compact set.

complementary function: The solution to the homogeneous part of a difference or differential equation.

complementary slackness: Complementary slackness refers to conditions that hold true between primal and dual problems at their optimal solution.

1. If the jth variable is positive in the optimal solution to the primal problem then the jth dual constraint holds as a strict inequality.

2. If the ith constraint in the primal problem holds as a strict inequality at the optimum then the ith dual variable is zero and is known as complementary slackness.

complex numbers: Numbers that include a real and imaginary part. Thus, $1 + 2i$ is a complex number where i is the imaginary number defined as $\sqrt{-1}$.

complex roots: Roots to a polynomial equation that are complex numbers.

composite function: a function whose argument is a function such that $g(c(x))$ is a composite function.

concave function: A function $f(x_1, \ldots, x_n)$ is concave if for all $\mathbf{x} = x_1, \ldots, x_n$ in the domain of the function (D), it is the case that

$$\lambda \cdot f(\mathbf{x}) + (1 - \lambda) \cdot f(\mathbf{y}) \leq f(\lambda \cdot \mathbf{x} + (1 - \lambda) \cdot \mathbf{y}),$$

where $x, y \in D$ and $\lambda \in [0,1]$. If the inequality holds strictly for all $\lambda \in (0,1)$ when $x \neq y$ then the function is strictly concave.

concave programming: This refers to a set of conditions in nonlinear programming which if satisfied along with the constraint qualification ensures that the Kuhn-Tucker conditions are both necessary and sufficient. The concave programming conditions are a stronger set of conditions than the Arrow-Enthoven conditions.

constraint qualification condition: In constrained optimization problems where x are the choice variables and $g_i(x)$ are the constraint functions and c_i are the constraints, the constraint qualification is satisfied if the objective function and constraint functions are differentiable and defined over a convex set and all the constraints are convex there also exists a non-zero vector $\underline{x}$ such that $g_i(x_1, \cdots, x_n) < c_i$ for all i **or** if the matrix of first-order partial derivatives of all the binding constraints evaluated at the maximum x^* has maximum rank. If all the constraints are linear the constraint qualification is automatically satisfied.

continuous function: A function is continuous at a point z if:

1. z is in the domain of the function and, hence, is defined at z.

2. $\lim_{x \to z} f(x)$ exists.

3. $\lim_{x \to z} f(x) = f(z)$.

continuously differentiable function: A function $f(x_1, x_2, \cdots, x_n)$ is continuously differentiable n times if **all** the partial derivatives exist up to n times and are continuous.

convergence: If $f(x) \to z$ as $x \to \infty$ where z is finite valued then $f(x)$ is said to converge to z. Variables which are a function of time are said to be convergent if $x(t) \to s$ as $t \to \infty$ where s is the stationary state.

convex function: A function $f(x_1, \ldots, x_n)$ is convex if for all $\mathbf{x} = x_1, \ldots, x_n$ in the domain of the function (D), it is the case that

$$\lambda \cdot f(\mathbf{x}) + (1 - \lambda) \cdot f(\mathbf{y}) \geq f(\lambda \cdot \mathbf{x} + (1 - \lambda) \cdot \mathbf{y}),$$

where $x, y \in D$ and $\lambda \in [0,1]$. If the inequality holds strictly for all $\lambda \in (0,1)$ when $x \neq y$ then the function is strictly convex.

convex set: A set is convex if for any two points x and y in the set then $\lambda x + (1 - \lambda)y \in S$ where S is the set and $\lambda \in (0, 1)$. This means that if we draw a line between any two points in the set, all points on the line remain within the set.

Cramer's rule : Cramer's rule is a method for solving for the unknowns in a system of linear simultaneous equations defined as $\mathbf{Ax} = \mathbf{b}$ and requires that the number of unknowns equal the number of equations. Provided that the inverse of the matrix of technical coefficients exists then the rule states that the ith element of the vector of unknowns $\mathbf{x}$ is $\mathbf{x_i} = \frac{\det \mathbf{B_i}}{\det \mathbf{A}}$ where the matrix $\mathbf{B_i}$ is the matrix $\mathbf{A}$ with the ith column replaced by the vector $\mathbf{b}$.

critical point: Also known as a stationary point and is where a function is neither increasing or decreasing.

cycling: A problem that is sometimes encountered in linear programming when there is degeneracy such that solution algorithm cycles between the same set of basic solutions without moving in the solution space.

decreasing function: For a function $f(x)$ where $a < b$ if $f(a) \geq f(b)$ then $f(x)$ is decreasing.

definite integrals: Integrals which have a numerical value and require an upper and lower limit of integration.

definiteness: The concept of definiteness is intimately related to quadratic forms and may be used to check the characteristics of functions. For example, if the Hessian of a function is positive (negative) definite over the domain of the function it is a convex (concave) function in its domain.

degeneracy: Degeneracy arises from cycling in the simplex method of linear programming from one feasible solution to another without reaching the optimum.

degree of a differential equation: The power to which the highest order derivative or differential is raised.

De Moivre's Theorem: A formula that is useful when working with complex numbers which states:

$$(a \pm bi)^n = [r(\cos\theta \pm i\sin\theta)]^n$$

dependent variable: A variable uniquely determined by a function.

derivative: The derivative of a function is the rate of change of the function with respect to its argument.

derivative function: The derivative of a primitive function. Thus, $f'(x)$ is the derivative function of $f(x)$.

determinant: Defined only for square matrices and is the sum of the product of the elements of any row or column multiplied by their respective cofactors. The determinant of a (2×2) matrix is the product of the elements of the main diagonal less the product of the other two elements. The determinant of a matrix $\mathbf{A}$ may be written as $|\mathbf{A}|$ or $\det(\mathbf{A})$.

difference equation: An equation that expresses changes in variables in discrete time. A solution to a difference equation contains no lagged terms and is consistent with the original equation.

difference quotient: The formal definition of the derivative can be defined in terms of the difference quotient as:

$$\frac{df(x)}{dx} = f'(x) = \lim_{\Delta x \to 0} \frac{f(z + \Delta x) - f(z)}{\Delta x}$$

where $\frac{df(x)}{dx}$ exists at $x = z$ if and only if the limit exists as $\Delta x \to 0$.

differentiable function: A function is differentiable at a point z if it is continuous and if the slope of the function to the left of z is equal to the slope of the function to the right of z, where z can be any point.

differential: For a function $y = f(x)$ we can denote a change in x and y by the differential of x (dx) and the differential of y (dy). Thus, the differential of the function $y = f(x)$ is

$$dy = f'(x)dx$$

differential equation: An equation that expresses the rates of change of variables in continuous time. A solution to a differential equation contains no derivative terms and is consistent with the original equation.

differentiation: The method of determining the derivative of a function. The important rules of differentiation include:

1. Power rule;

$$\frac{dx^n}{dx} = nx^{n-1}$$

2. Chain rule for composite functions;

$$\frac{df(g(x))}{dx} = \frac{df(g(x))}{dg(x)} \cdot \frac{dg(x)}{dx}.$$

3. Product rule:

$$\frac{d(f(x) \cdot g(x))}{dx} = g(x) \cdot \frac{df(x)}{dx} + f(x) \cdot \frac{dg(x)}{dx}.$$

4. Quotient rule;

$$\frac{d}{dx}\frac{f(x)}{g(x)} = \frac{\frac{df(x)}{dx}g(x) - f(x)\frac{dg(x)}{dx}}{(g(x))^2}$$

5. Logarithmic rule;

$$\frac{d\ln(x)}{dx} = \frac{1}{x}$$

6. Exponential rule;

$$\frac{de^{ax}}{dx} = ae^x$$

dimension: The number of columns and rows of a matrix.

distance formula: Defines the distance between two points (x_1, y_1) and (x_2, y_2) as the square root of $(x_1 - x_2)^2 + (y_1 - y_2)^2$.

distinct real roots: Refer to the roots of a polynomial equation which are all unique real numbers.

divergence: The opposite of convergence.

domain: For the function defined as $y = f(x)$, the domain of the function is the set of all permissible values of x.

duality: A way of describing optimization problems in another way. For example, if the primal problem is to maximize revenue subject to a set of contraints then the dual problem is to minimize the resource cost of the scarce resources.

eigenvalues: Also called characteristic roots, are the roots to a characteristic equation.

eigenvectors: The expression obtained from substituting into the characteristic equation the characteristic roots or eigenvalues.

elementary row operations: Row operations that include:

1. Interchange of rows.
2. Mutiplying all the elements in a row by a non-zero scalar.
3. Rows and constant multiple of rows can be added or subtracted.

endogenous variable: A variable that depends on the equation, system of equations or model.

Envelope Theorem: It states that the partial derivative of the optimal value function with respect to a parameter equals the partial derivative of the direct objective function or the associated Lagrangean with respect to the same parameter, holding all variables at their optimal values.

equilibrium point: The derived values of the endogenous variables in terms of exogenous variables or parameters of a system of equations or economic model.

equilibrium state: Also known as a stationary state and almost always refers to variables which are a function of time. A stationary state is where the value of the variable is unchanging with respect to time.

Euclidean formula: The length of a vector $\mathbf{x} \in E^n$ can be computed by the formula:

$$|\mathbf{x}| = \sqrt{x_1^2 + x_2^2 + \cdots x_n^2}$$

where n is the dimension of the vector.

Euclidean space: Denoted by E^n is a vector space of vectors on n components which is closed under addition and scalar multiplication.

Euler's Theorem: This theorem states that if a function $f(x_1, \ldots, x_n)$ is homogeneous of degree k then

$$\frac{\partial f}{\partial x_1} x_1 + \frac{\partial f}{\partial x_2} x_2 \ldots \frac{\partial f}{\partial x_n} x_n \equiv k f(x_1, x_2, \ldots x_n)$$

exact differential equation: A partial differential equation with two variables (x, y) is exact if the total differential of the primitive function $F(x, y)$ is zero.

exogenous variable: A variable that is independent of the equation, system of equations or model and used to determine the values of the endogenous variables.

exponent: The power to which a variable is raised is called the exponent of that variable.

exponential function: An exponential function is written as

$$f(x) \;=\; Ab^x$$

where A is a constant, b is the base of the function and x is the exponent.

factorial: n factorial is written as $n!$ where n is a positive integer and equals $1 \times 2 \times \cdots \times (n-1) \times n$.

first-order conditions: Often called necessary conditions and must be satisfied for the function in question to have a critical point.

full rank: A square matrix has full rank if the rank equals the order of the matrix itself.

function: A set of ordered pairs which may be defined as $y = f(x)$ where every x determines a unique y value.

geometric series: A series of numbers defined as follows $a + ac + ac^2 + ac^3 \ldots ac^n$. The sum of an infinite geometric series is $\frac{a}{1-c}$ where $|c| < 1$. The sum of a finite geometric series with n terms is $a\frac{1-c^n}{1-c}$ where $c \neq 1$.

Hawkins-Simon Conditions: The Hawkins-Simon conditions require that the Leontief matrix $(\mathbf{I} - \mathbf{A})$ in input-output analysis have the following characteristics:

1. All the elements on the main diagonal must be positive.

2. All the leading principal minors must be positive.

Hessian matrix: The matrix of second-order partial derivatives of a function.

homogeneity: A function $f(x)$ is homogeneous of degree k if $f(\lambda x) = \lambda^k f(x)$ for all $\lambda > 0$.

homogeneous difference equation: A difference equation is homogeneous if it contains only lagged variables and no constant term.

homogeneous differential equation: A differential equation is homogeneous if it contains terms that only include the dependent variable or its derivatives and no constant term.

homogeneous system: A system of linear equations defined by $\mathbf{Ax} = \mathbf{b}$ where $\mathbf{b}$ is a null column vector which consists entirely of zeros is called a homogeneous system.

homothetic functions: These are functions which are, in general, not homogeneous but are a monotonic transformation of a homogeneous function. Thus, if $f(x_1, x_2)$ is homogeneous of degree k provided that $V'(f(x_1, x_2)) > 0$ then the function $V(f(x_1, x_2))$ is homothetic. It follows that homogeneous functions are a subset of homothetic functions. The level sets of a homogeneous function are radial blowups of each other.

Hotelling's Lemma: Used to obtain the input demand and output supply functions from a profit function. It states that if $\Pi(p, w)$ is the profit function where p and w are, respectively, the output price and vector of input prices then the output supply is $\frac{\partial \Pi(p,w)}{\partial p}$ and the input demand i is $\frac{-\partial \Pi(p,w)}{\partial w_i}$.

idempotent matrix: A matrix which equals the product of the matrix multiplied by itself. The identity matrix is an idempotent matrix because $\mathbf{I_n I_n} = \mathbf{I_n}$.

identity matrix: A matrix whose elements along the main diagonal are all equal to one and all other elements are equal to zero. Post or premultiplying a matrix $\mathbf{A}$ by its inverse results in the identity matrix.

implicit function: A function where the dependent variable is not written explicitly in terms of the independent variables.

implicit function theorem: This theorem permits us to solve the derivatives of a continuous differentiable function that cannot be written explicitly. The theorem states that given $F(y, x) = 0$ where x is a vector and if the implicit function $y = f(x_1, x_2 \ldots x_n)$ exists and is continuously differentiable then its partial derivatives are:

$$\frac{\partial y}{\partial x_i} = -\frac{\frac{\partial F}{\partial x_i}}{\frac{\partial F}{\partial y}}$$

provided that $\frac{\partial F}{\partial y} \neq 0$.

improper integrals: Integrals that arise when one of the limits of integration is infinite.

inconsistency: If there exists no solution to a system of equations we call that system inconsistent.

increasing function: For a function $f(x)$ where $a < b$ if $f(a) \leq f(b)$ then $f(x)$ is increasing.

indefinite: A matrix which is neither positive nor negative semidefinite is said to be indefinite.

indefinite integrals: Integrals for which the solution is a primitive function that has no specific numerical value.

independent variable: A variable independent of a functional relationship but which helps determine the value of dependent variables.

indirect objective function: see optimal value function.

infeasibility: Infeasibility arises when it is impossible to satisfy the constraints in an optimization problem.

inflection point: A point in the domain of the function where the second-order derivative is zero.

input-output analysis: A widely-used technique that is applied, among other things, to examine demand and supply relationships in various sectors and industries of an economy at a particular point in time.

integers: The set of integers includes $\cdots -2, -1, 0, 1, 2 \cdots$.

integration: The process by which a primitive function is obtained from a derived function. The main rules of integration where $f(x)$ is defined as the integrand and is K is a constant of integration are:

1. Multiplication by a constant rule;

$$\int cf(x)dx = c\int f(x)dx$$

   where c is any real number

2. Power rule;

$$\int x^a dx = \frac{1}{a+1}x^{a+1} + K$$

   where $a \neq -1$.

3. Exponential rule;

$$\int e^x dx = e^x + K$$

4. Logarithmic rule;

$$\int \frac{f'(x)}{f(x)}dx = \ln|f(x)| + K$$

   where $f(x) \neq 0$.

5. Substitution rule;

$$\int f(u)\frac{du}{dx}dx = \int f(u)du$$

6. Integration by parts rule;

$$\int u(x)v'(x)dx = u(x)v(x) - \int v(x)u'(x)dx$$

integrating factor: A factor used to make a partial differential equation exact.

interior: The interior of a bordered Hessian matrix, not including the border.

intermediate value theorem: The theorem states that if a continuous function $g(x)$ takes on the value $g(a)$ at a, and $g(b)$ at b where $g(a) < 0 < g(b)$ in a closed interval $[a, b]$ then there exists a point c in the open interval (a, b) such that $g(c) = 0$. More generally, if $g(x)$ is continuous on $[a, b]$ then $g(x)$ must assume every value between $g(a)$ and $g(b)$ at some point x in the interval (a, b).

inverse function: Provided that a function $y = f(x)$ has a unique x for every y—a one-to-one mapping of x on y—the inverse function $x = f^{-1}(y)$ exists.

inverse matrix: The inverse of a square matrix $\mathbf{A}$ is defined as $\mathbf{A}^{-1}$ and when pre or post multiplied by $\mathbf{A}$ yields the identity matrix.

inverse method: A method for solving a system of linear simultaneous equations where the number of unknowns equals the number of equations. If the system of equations is defined as $\mathbf{Ax} = \mathbf{b}$ where $\mathbf{A}$ is the matrix of coefficients, $\mathbf{x}$ is the vector of unknowns and $\mathbf{b}$ is a vector of known constants then the solution via the method of the inverse is $\mathbf{x} = \mathbf{A}^{-1}\mathbf{b}$ provided that $\det(A) \neq 0$.

invertible matrix: Any square matrix where the determinant does not equal zero is called an invertible or a non-singular matrix.

irrational numbers: Irrational numbers are those real numbers that cannot be formed as a ratio of integers such as $\sqrt{2}$.

Jacobian: A matrix that can be used to test for functional dependence among n, not necessarily linear, equations and is the matrix of the first order partial derivatives of the system. For example, for the system:

$$F_1(x_1, x_2, y_1, y_2) = 0$$
$$F_2(x_1, x_2, y_1, y_2) = 0$$

where x_1 and x_2 are the exogenous variables treated as given and y_1 and y_2 are the endogenous variables for which we must solve. The Jacobian matrix is

$$\mathbf{J} = \frac{\partial \mathbf{F}}{\partial \mathbf{y}} = \begin{pmatrix} \partial F_1 / \partial y_1 & \partial F_1 / \partial y_2 \\ \partial F_2 / \partial y_1 & \partial F_2 / \partial y_2 \end{pmatrix}.$$

Jacobian determinant: A test for functional dependence among the n equations is referred to as the Jacobian determinant, which is defined as $\det(\mathbf{J})$ or $|\mathbf{J}|$. If the determinant equals zero for all values of y then the equations are functionally dependent and it is not possible to define the endogenous variables solely in terms of the exogenous variables. Conversely, if $|\mathbf{J}| \neq 0$ such that the Jacobian is non-singular and if each $\mathbf{F}_i$ is continuously differentiable then it is possible to define the endogenous variables in terms of the exogenous variables in a neighbourhood of the equilibrium point (x_0, y_0) which satisfies the original system of equations.

Kuhn-Tucker conditions: A set of conditions for nonlinear programming problems which may be used to find an optimum. If the constraint qualification condition is satisfied, the Kuhn-Tucker conditions are necessary and if the Arrow-Enthoven conditions are also satisfied they are both necessary and sufficient.

Lagrangean function: The direct objective function plus each of the constraints appropriately written multiplied by a Lagrangean multiplier.

Lagrangean multiplier: A variable created when the Lagrangean function is formed that is often treated as a choice variable when obtaining the first-order conditions. A positive value for a Lagarangean multiplier in a maximization problem indicates the associated constraint is binding.

leading principal minors: The determinants of the submatrices formed along the main diagonal of a matrix.

left hand limit: The limit of a function $f(x)$ as we approach a number z from a lower value, i.e., $\lim_{x \to z^-} f(x)$.

Leontief matrix: The Leontief matrix is defined as $(\mathbf{I} - \mathbf{A})$ in input-output analysis where $\mathbf{I}$ is the identity matrix and $\mathbf{A}$ is the matrix of technical coefficients.

level sets: the combinations of the variables of a function which give the same function values. The level sets of a utility function, production function and cost function are called indifference curves, isoquants and isocost curves.

L'Hôpital's Rule: If two functions $f(x)$ and $g(x)$ are differentiable in an interval about a point a but not necessarily at a and if $f(x) \to 0$ and $g(x) \to 0$ as x approaches a then:

$$\lim_{x \to a} \frac{f(x)}{g(x)} = \lim_{x \to a} \frac{f'(x)}{g'(x)}$$

provided that $g'(x) \neq 0$ for all $x \neq a$.

limit: The value of a function as its argument approaches some number z, i.e., $\lim_{x \to z} f(x)$. Some useful rules when solving the limits of a function are listed below where K is a constant and $\lim_{x \to z} f(x) = A$ and $\lim_{x \to z} g(x) = B$.

1. $\lim_{x \to z} K f(x) = KA$.
2. $\lim_{x \to z} [f(x) \pm g(x)] = A \pm B$.
3. $\lim_{x \to z} f(x) g(x) = AB$.
4. $\lim_{x \to z} \frac{f(x)}{g(x)} = \frac{A}{B}$.
5. $\lim_{x \to z} [f(x)]^j = A^j$.
6. If $f(x)$ and $g(x)$ are equal for all values of x close to z but not necessarily at z then $\lim_{x \to z} f(x) = \lim_{x \to z} g(x)$ whenever the limit exists.

linear difference equation: A difference equation is said to be linear if no variable is raised to a power greater than one and is not multiplied by any other term of another period.

linear differential equation: A differential equation is said to be linear if no variable, derivative or differential term is raised to a power greater than one and there is no derivative or differential term that is multiplied by a variable.

linear function and equation: A linear function is a first degree polynomial defined as $a_0 + a_1 x$ where $a_1 \neq 0$. A linear equation is formed from a linear function and is defined as $a_0 + a_1 x = 0$ where $a_1 \neq 0$.

linear independence: A system of linear equations is said to be linearly independent if it is not possible to express any one of the equations as a linear combination of the other equations.

linear programming: Refers to optimization problems where the objective function and all the constraints are linear.

logarithm: A logarithm is the power to which a base must be raised to obtain a particular number. For example, $\log_3 9 = 2$ or the base 3 raised to the power of 2 equals 9.

logarithmic function: The logarithmic function and exponential function are intimately related in that if $y = e^x$ is the exponential function then $y = \log_e x$ is its inverse and is the logarithmic function. A logarithmic function is a function where the dependent variable is expressed as the logarithm of the independent variable.

lower triangular matrix: A lower triangular matrix is a square matrix where all the elements to the right of the main diagonal are zero.

main diagonal: The main diagonal of a square matrix are the elements formed by drawing a line from the upper left corner to the bottom right corner.

matrix: A matrix is a systematic arrangement of elements into rows and columns.

maximum: A point x^* is a local maximum if:

$$f(x^*) \geq f(x) \text{ for all } x \in (x^* - \epsilon, x^* + \epsilon),$$

where $\epsilon > 0$ is as small as necessary. A point x^* is a global maximum if:

$$f(x^*) \geq f(x) \text{ for all } x$$

maximum rank: The maximum rank of a matrix is the minimum of [number of rows, number of columns].

minimum: A point x^* is a local minimum if:

$$f(x^*) \leq f(x) \text{ for all } x \in (x^* - \epsilon, x^* + \epsilon),$$

where $\epsilon > 0$ is as small as necessary. A point x^* is a global minimum if:

$$f(x^*) \leq f(x) \text{ for all } x$$

minor: The minor of an element a_{ij} is defined as the determinant of the matrix formed by deleting the ith row and jth column.

monotonicity: Functions that are strictly increasing (decreasing) are sometimes described as monotonically increasing (decreasing).

multiple integral: When integrating functions of several variables with respect to each of the variables, the problem is called a repeated or multiple integral.

necessary conditions: Often known as the first-order conditions, are the conditions that must be satisfied if there is a maximum or minimum of a function. More generally, necessary conditions are conditions that must be satisfied to ensure a particular outcome such as an equilibrium or a solution to an optimization problem.

nonlinear programming: Optimization problems where either the objective function and/or one or more of the constraints are not linear in the choice variables.

non-singular: A matrix is non-singular if the inverse of the matrix exists such that its determinant is non-zero.

non-trivial solution: A homogenous system of linear equations is said to have a non-trivial solution when the coefficient matrix is singular, so that $\det(\mathbf{A}) = 0$.

normalization: A vector can be normalized to be of unit length by ensuring that the sum of the square of each element in the vector equals 1, i.e., if $\mathbf{v} = (a_1, a_2, \ldots a_n)$ then $\sum_{i=1}^{n} a_i^2 = 1$.

Nth derivative test: A test to determine local maxmima and minima when a single variable function $f(x)$ has a second derivative equal to zero. It states that if x^* is a stationary point of an Nth differentiable function and if $f^n(x^*) < 0$ and n is an even number then x^* is a local maximum and if n is an odd number it is a point of inflection, and if $f^n(x^*) > 0$ and n is an odd number then x^* is a local minimum.

objective function: Also called the direct objective function which must be maximized or minimized by choosing the optimal values of the choice variables.

open set: A set which does not contain any of its boundary points. The set $(0, 1)$ is an open set as it contains all values between 0 and 1 but not the boundary points, 0 and 1.

optimal value function: A function only of the parameters in the constraints and direct objective function. The optimal value function is sometimes known as the maximum or minimum value function or the indirect objective function and is formed by substituting into the direct objective function the optimal values of the choice variables.

order of difference and differential equations: The order of a difference equation refers to the highest number of periods lagged. The order of a differential equation is the highest order of the derivatives appearing in the equation.

ordinary differential equation: A differential equation which has only one dependent variable.

orthogonal: Two vectors are orthogonal if the vectors are at right angles or perpendicular to each other.

orthonormal: A set of vectors which are mutually orthogonal and each is of unit length is an orthonormal set.

parameters: Values given in an economic model, problem, or system of equations. For example, in utility maximization problems the consumer's budget and the price of commodities may be considered to be parameters.

partial derivative: The change in the dependent variable (y) at a point from a change in one of the independent variables (x_i), holding all other variables fixed, is called the partial derivative and is denoted by $\frac{\partial y}{\partial x_i}$.

partial differential equation: If the solution or primitive function is a function of more than one variable then we have a partial differential equation.

partial differentiation: The method for determining the partial derivatives of a function.

particular integral or solution: The so-called intertemporal equilibrium of a differential or difference equation and is any solution that satisfies the equation.

phase diagram: A diagram that represents a qualitatitive solution of an autonomous differential equation.

polynomial function and equation: The polynomial function has the following form:

$$P(x) = a_0 + a_1 x + a_2 x^2 + \ldots + a_n x^n$$

where each a_i term is a real number while each x term is raised to a non-negative integer. If $a_n \neq 0$ the polynomial is of degree n. The degree of the polynomial refers to the highest power to which the variable is raised. A polynomial equation formed from a function of degree n is defined as:

$$a_0 + a_1 x + a_2 x^2 + \ldots + a_n x^n = 0$$

where $a_n \neq 0$.

primitive function: The integral of the derivative function.

principal minors: The determinants of the submatrices formed by deleting k columns and rows from a matrix where $k < n$ and n is the dimension of the matrix.

quadratic form: A quadratic form is a function defined as

$$Q(x) = a_{11}x_1 + a_{12}x_1 x_2 + \ldots + a_{ij}x_{ij} + \ldots + a_{nn}x_{nn}^2$$

The quadratic form may also be defined using matrices such that $Q(x) = \mathbf{x}'\mathbf{A}\mathbf{x}$ where A is a square symmetric matrix. If $Q(x)$ is always positive (negative) for all values of x where $(x_1, x_2 \ldots x_n) \neq 0$ then the matrix $\mathbf{A}$ is positive (negative) definite. If $Q(x)$ is always equal to or greater (less) than 0 for all values of x where $(x_1, x_2 \ldots x_n) \neq 0$ then the matrix $\mathbf{A}$ is positive (negative) semidefinite.

quadratic formula: Used to solve quadratic equations where if $ax^2 + bx + c = 0$ then if $a \neq 0$

$$x = \frac{-b \pm \sqrt{b^2 - 4ac}}{2a}$$

The formula may yield two, one or possibly no values of x that are real numbers.

quadratic function and equation: A quadratic function is a second degree polynomial of the form $a_0 + a_1 x + a_2 x^2$ where $a_2 \neq 0$. A quadratic equation is formed from a quadratic function and is defined as $a_0 + a_1 x + a_2 x^2 = 0$ where $a_2 \neq 0$.

quadratic programming: Optimization problems which are to be solved using a modified form of the simplex method provided that the objective function involves a quadratic function and is strictly concave and all the constraints are linear.

quasiconcave programming: Nonlinear programming problems that satisfy the Arrow-Enthoven conditions.

quasiconcavity: A function $f(x)$ is said to be quasiconcave if for any two values of x defined by (x_1, x_2) where $f(x_1) = c$ and $f(x_2) \geq c$, it is also true that

$$f(\lambda x_1 + (1 - \lambda)x_2) \geq c$$

for all $\lambda \in [0,1]$. An alternative definition of quasiconcavity is that the upper contour set sometimes called the weakly better than set in economics is convex, i.e., the set S

$$S = \{x : f(x) \geq c\},$$

is convex where c is an arbitrary constant. All concave functions are quasiconcave but the reverse is not true.

quasiconvexity: A function $f(x)$ is said to be quasiconvex if for any two values of x defined by (x_1, x_2) where $f(x_1) = c$ and $f(x_2) \leq c$, it is also true that

$$f(\lambda x_1 + (1 - \lambda)x_2) \leq c$$

for all $\lambda \in [0,1]$. A function is strictly quasiconvex if the above expression holds as a strict inequality and $\lambda \in (0, 1)$. An alternative definition of quasiconvexity is that the lower contour set, sometimes known as the weakly worse than set in economics, is convex, i.e., the set S

$$S = \{x : f(x) \leq c\},$$

is convex where c is an arbitrary constant. All convex functions are quasiconvex but the reverse is not true.

range: The set of all possible values of a function.

rank: The rank of a matrix equals the number of linearly independent columns. The rank of any matrix equals the order of the largest square matrix which can be formed by deleting rows, columns, or both, which has a non-zero determinant.

rational numbers: All numbers formed by any ratio of integers defined as $\frac{a}{b}$ where $b \neq 0$. Thus, $\frac{-1}{2}$ is a rational number but the $\sqrt{2}$ is not.

real numbers: Real numbers include all rational and irrational numbers.

reduced matrix: A reduced matrix has the following characteristics:

1. Provided that the row does not contain all zeros, the first or leading element must equal 1.

2. The leading element of each row is to the right of the leading element of the row above.

3. All zero rows (where all elements are zero) are at the bottom of the matrix.

reduction method: The method of reduction uses elementary row operations to derive a reduced matrix from the augmented coefficient matrix defined as $[\mathbf{A}|\mathbf{b}]$ for a system of simultaneous equations.

relation: A set of ordered pairs (x_i, y_j) where $i = i, 2 \ldots n$ and $j = 1, 2, \ldots m$.

resource cost: The quantity of the resources used multiplied by their shadow prices.

right hand limit: The right hand limit is what happens to the value of a function $f(x)$ as we approach a number z from a higher value, i.e., $\lim_{x \to z^+} f(x)$.

roots: The values that solve a polynomial equation. The roots of a quadratic equation are found by using the quadratic formula.

Routh-Hurwitz Theorem: A method for determining whether the solution to higher order differential equations converges or diverges.

Roy's Identity: Used to obtain the Marshallian demand functions from the indirect objective function. If x_i is the Marshallian demand for good i, $V(p, m)$ is the indirect utility function, p is a vector of prices of goods and m is the consumer's budget then $x_i = -\frac{\partial v(p,m)}{\partial p_i} / \frac{\partial v(p,m)}{\partial m}$.

saddle-point: A point on the surface of a function which is neither a local maximum nor minimum but which is nevertheless flat in every direction. For multivariable functions if the determinant of the Hessian matrix does not equal zero and the leading principal minors do not satisfy the condition for positive or negative definiteness then $\mathbf{x}^*$ satisfies a sufficient condition to be a saddle-point.

scalar: A single number rather than a vector or matrix.

Schur's Theorem: A theorem used to determine whether the solution to higher order difference equations converges or diverges.

second derivative test: A test for a local maximum or minimum of one variable functions. A stationary point x^* is a local maximum if $\frac{d^2 f(x^*)}{dx^2} < 0$ and is a local minimum if $\frac{d^2 f(x^*)}{dx^2} > 0$.

second-order conditions: In optimization problems, the conditions which ensure the critical point found with the first-order conditions is indeed a maximum or a minimum. The conditions are sufficient such that failure to satisfy the second-order conditions does not necessarily mean that the critical point is not a maximum or a minimum.

semidefiniteness: There are two types—positive semidefiniteness and negative semidefiniteness. For negative semidefiniteness all the principal minors of order n obtained from a matrix of order n defined as $|\mathbf{B_n}|$ must satisfy $(-1)^n |\mathbf{B_n}| \geq 0$ while for positive semidefiniteness all the principal minors ≥ 0.

separable differential equation: A differential equation is separable if the independent and dependent variables can be expressed as separate functions, i.e.,

$$\frac{dx}{dt} = f(x)g(t)$$

sequence: A sequence is a list of numbers. An infinite sequence of real numbers $a_0, a_1, a_2 \ldots$ converges to a limiting value C if for any ϵ, not matter how small, the sequence contains a term a_n such that all terms in the sequence following a_n have the property: $|a_m - C| < \epsilon$ where $m > n$. A sequence which does not converge to a real number is said to diverge.

series: A series is the sum of a list of numbers. An infinite series is $a_0 + a_1 + a_2 \ldots$. If the sum of an infinite series is a finite number then the series is said to be convergent.

set: A set is a collection of objects whose members are called elements.

shadow price: The rate at which the optimal value of the objective function changes with respect to a change in a constraint.

Shephard's Lemma: Used to obtain the input demand functions from a cost or expenditure function. It states that if $C(w, y)$ is the cost function where w is a vector of input prices and y is output then demand for input i is $\frac{\partial C(w, y)}{\partial w_i}$.

simplex method: An algorithm for solving linear programming problems.

singular matrix: A matrix whose determinant is zero.

slack variables: Slack variables are those variables which are added to $\leq$ constraints in linear programming problems. For example, the constraint $2x_1 + 3x_2 \leq 10$ becomes $2x_1 + 3x_2 + s_1 = 10$ where s_1 is a slack variable.

span: A set is said to span an n dimensional Euclidean space if every vector with n elements can be written as a linear combination of the set.

square matrix: A square matrix is a matrix where the number of rows equals the number of columns.

stability: The time path of the variable is said to be stable if the variable converges to the particular solution or stationary state over time.

standard form: A method of writing a linear programming problem so as to apply the simplex method to obtain the optimum.

stationary point: Also known as a critical point, is a point where the function is neither increasing nor decreasing. If the function is differentiable it corresponds to a point where the first derivative is zero.

stationary state: Sometimes called an equilibrium state and almost always refers to variables which are a function of time. A stationary state is where the value of the variable is unchanging with respect to time.

strictly decreasing function: For a function $f(x)$ where $a < b$ if $f(a) > f(b)$ then $f(x)$ is strictly decreasing.

strictly increasing function: For a function $f(x)$ where $a < b$ if $f(a) < f(b)$ then $f(x)$ is strictly increasing.

structural variables: The original choice variables in a linear programming problem.

sufficient conditions: In general, the conditions which if true ensure a given outcome. Sufficient conditions are not necessary in that the outcome may still arise even if the sufficient conditions are not satisfied.

surplus variables: Surplus variables are those which are substracted from $\geq$ constraints in linear programming. For example, the constraint $2x_1 + 3x_2 \geq 10$ becomes $2x_1 + 3x_2 - s_1 = 10$ where s_1 is a surplus variable.

symmetric matrix: A symmetric matrix is a square matrix where $a_{ij} = a_{ji}$ for $i \neq j$.

Taylor series: This involves approximating the value of a function by the use of a polynomial around a point. The Taylor series approximation of the function $f(x)$ around the point x_0 is:

$$f(x) \approx f_n(x) = f(x_0) + f'(x_0)(x - x_0) + \ldots + \frac{f^n(x_0)}{n!}(x - x_0)^n$$

where ! denotes factorial and $n!$ is defined as $1 \times 2 \times 3 \ldots n$.

technical coefficients: The elements in the input-output table are called technical coefficients.

total differential: In the case where the function has two or more independent variables we can apply the notion of a total differential. For example, the total differential of $y = f(x_1, x_2, \ldots, x_n)$ is:

$$dy = \frac{\partial f(x)}{\partial x_1}dx_1 + \frac{\partial f(x)}{\partial x_2}dx_2 + \ldots + \frac{\partial f(x)}{\partial x_n}dx_n$$

total differentiation: The method to obtain the total differential of a function.

transpose: A transpose of a matrix $\mathbf{A}$ is defined as $\mathbf{A}^{\mathbf{T}}$ or $\mathbf{A}'$ and is formed by interchanging the rows and columns.

unboundedness: Refers to optimization problems where the constraints do not prevent the possibility of a solution that is not finite valued.

upper triangular matrix: An upper triangular matrix is a square matrix where all the elements to the left of the main diagonal equal zero.

variable: A symbol, often denoted by a letter, that represents something that can vary.

vector: A matrix consisting only of one row or column.

vector space: Any set of vectors which is closed under addition and scalar multiplication such that if $x_1, x_2, \ldots, x_n$ are vectors then the set of all linear combinations $\alpha_1 x_1 + \alpha_2 x_2 + \ldots + \alpha_n x_n$ is a vector space.

Young's Theorem: If a function is twice continuously differentiable then the cross-partial derivatives are invariant to the order of differentiation. Thus, for a twice continuously differentiable function $f(x_1, x_2)$ it is the case that $f_{12} = f_{21}$.

Bibliography

[1] Archibald, G.C. and R.G. Lipsey. 1977. *An Introduction to a Mathematical Treatment of Economics*, Third Edition, Weidenfeld and Nicolson: London.

[2] Baldani. J., J. Bradfield and R. Turner. 1996. *Mathematical Economics* Dryden Press, Harcourt Brace College Publishers: New York.

[3] Baumol, W.G. 1970. *Economic Dynamics: An Introduction*, Third Edition, Macmillan: London.

[4] Beavis, B. and I.M. Dobbs. 1990. *Optmization and Stability Theory for Economic Analysis*, Cambridge University Press: New York.

[5] Binmore, K.G. 1983. *Calculus*, Cambridge University Press: Cambridge, England.

[6] Birchenhall, C. and P. Grout. 1984. *Mathematics for Modern Economics* Barnes and Noble Books: Totawa, New Jersey.

[7] Black, J. and J.F. Bradley. 1980. *Essential Mathematics for Economists*, Second Edition, John Wiley and Sons: Chichester, England.

[8] Blackorby, C., D. Primont and R.R. Russell. 1978. *Duality, Separability and Functional Structure*, North-Holland: New York.

[9] Bradley, S.P., A.C. Hax, and T.L. Magnanti. 1977. *Applied Mathematical Programming*, Addison-Wesley: Reading, Mass.

[10] Bressler, B. 1975. *A Unified Introduction to Mathematical Economics*, Harper and Row: New York.

[11] Brock, W.A. and A.G. Malliaris. 1989. *Differential Equations, Stability and Chaos in Dynamic Economies*, Elsevier Science Publishing: Amsterdam.

[12] Chambers. R.G. 1988. *Applied Production Analysis: A Dual Approach*, Cambridge University Press: New York.

[13] Chiang, A.C. 1984. *Fundamental Methods of Mathematical Economics*, Third Edition, McGraw-Hill: Toronto.

[14] Deaton, A. and J. Muellbauer. 1983. *Economics and Consumer Behavior*, Cambridge University Press: New York.

[15] Dixit, A.K. 1990. *Optimisation in Economic Theory*, Second Edition, Oxford University Press: New York.

[16] Dorfman, R., P.A. Samuelson, and R.M. Solow. 1958. *Linear Programming and Economic Analysis*, McGraw-Hill: New York.

[17] Dowling, E.T. 1990. *Mathematics for Economists*, Second Edition, McGraw-Hill: New York.

[18] Gandolfo, G. 1971. *Mathematical Methods and Models in Economic Dynamics*, North-Holland: Amsterdam.

[19] Glaister, S. 1984. *Mathematical Methods for Economists*, Third Edition, Basil Blackwell: Oxford.

[20] Goldberg, S. 1958. *Introduction to Difference Equations*, John Wiley: New York.

[21] Haeussler E.F. Jnr. and R. Paul. 1993. *Introductory Mathematical Analysis for Business, Economics and Life and Social Sciences*, Seventh Edition, Prentice-Hall: Englewood Cliffs, New Jersey.

[22] Hands, D. W. 1991. *Introductory Mathematical Economics*, D.C. Heath and Company: Toronto.

[23] Holden, K. and A.W. Pearson. 1983. *Introductory Mathematics for Economists*, MacMillan: London.

[24] Lambert, P. 1985. *Advanced Mathematics for Economists*, Basil Blackwell: Oxford.

[25] Ostrosky, A.L. and J.V. Koch. 1979. *Introduction to Mathematical Economics*, Houghton Mifflin: Boston.

[26] Rowcroft, J.E. 1994. *Mathematical Economics: An Integrated Approach*, Prentice Hall Canada: Scarborough, Ontario.

[27] Rudin, W. 1976. *Principles in Mathematical Analysis*, Third Edition, McGraw-Hill: New York.

[28] Schrage, L. 1991. *LINDO An Optimization and Modelling System*, Fourth Edition. The Scientific Press: San Francisco.

[29] Silberberg, E. 1990. *The Structure of Economics: A Mathematical Analysis*, Second Edition, McGraw-Hill: New York.

[30] Strang, G. 1988. *Linear Algebra and its Applications*, Third Edition, Harcourt, Brace and Jovanich: San Diego.

[31] Sydsæter, K. 1981. *Topics in Mathematical Analysis for Economists*, Academic Press: London.

[32] Sydsæter, K. and P.J. Hammond. 1995. *Mathematics for Economic Analysis*, Prentice-Hall: Englewood Cliffs, New Jersey.

[33] Takayama, A. 1985. *Mathematical Economics*, Second Edition, Cambridge University Press: New York.

[34] Toumanoff, P. and F. Nourzad. 1994. *A Mathematical Approach to Economic Analysis*, West Publishing Company: Minneapolis.

[35] Varian H.R. 1993. *Microeconomic Analysis*, Third Edition, W.W. Norton and Company: New York.

[36] Wismer, D.A. and R. Chattergy. 1978. *Introduction to Nonlinear Optimization*, North-Holland: New York.

[37] Woods, J.E. 1978. *Mathematical Economics*, Longman: New York.

[38] Wu, N. and R. Coppins. 1981. *Linear Programming and Extensions*, McGraw-Hill: New York.

Index